COMBINATION PRODUCTS

Regulatory Challenges and Successful Product Development

COMBINATION PRODUCTS

Regulatory Challenges and Successful Product Development

Smita Gopalaswamy • Venky Gopalaswamy

CRC Press
Taylor & Francis Group
Boca Raton London New York

CRC Press is an imprint of the
Taylor & Francis Group, an **informa** business

CRC Press
Taylor & Francis Group
6000 Broken Sound Parkway NW, Suite 300
Boca Raton, FL 33487-2742

Printed in the United States of America on acid-free paper
10 9 8 7 6 5 4 3 2 1

International Standard Book Number-13: 978-1-4200-6446-9 (Hardcover)

Library of Congress Cataloging-in-Publication Data

Gopalaswamy, Smita.
Combination products : regulatory challenges and successful product developement / Smita Gopalaswamy and Venky Gopalaswamy.
p. ; cm.
Includes bibliographical references and index.
ISBN 978-1-4200-6446-9 (hardcover : alk. paper)
1. Drugs--Standards--United States. 2. Medical instruments and apparatus--Standards--United States. 3. Biological products--Standards--United States. 4. United States. Food and Drug Administration. I. Gopalaswamy, Venky. II. Title.
[DNLM: 1. United States. Food and Drug Administration. 2. Equipment Design--standards--United States. 3. Pharmaceutical Preparations--standards--United States. 4. Biological Products--standards--United States. 5. Device Approval--standards--United States. 6. Equipment and Supplies--standards--United States. QV 26 G659c 2008]

RM301.27.G68 2008
615'.1--dc22 2007043090

Visit the Taylor & Francis Web site at
http://www.taylorandfrancis.com

and the CRC Press Web site at
http://www.crcpress.com

Dedication

To my late parents, who always encouraged me to do my best, all my academic mentors, and colleagues in my professional journey.

—S. G.

To professionals in medical device, pharmaceutical, and biologics industries working to successfully develop and launch combination products.

—V. G.

Contents

About the authors

Dr. Smita Gopalaswamy is the president of Advanced Solutions Consultants, Inc. (ASC), a management consulting company to the medical device, pharmaceutical, and biologics industries. Prior to this role, she had strong experience in industry as a professional and also as a consultant, having worked with Johnson & Johnson, Mt. Sinai Medical Center, Boston Scientific, Ivax, Beckman Coulter, and so on.

Dr. Smita Gopalaswamy has more than 20 years of leadership and technical consulting responsibilities in the medical device, biologics, and pharmaceutical industries, and with combination products. Her quality, compliance, and regulatory expertise areas include FDA and ISO interactions and audits, submissions and consulting, CE (Conformité Europeénne) marking and device certification, quality systems development and deployment, the new product development process, and ISO, quality system regulations (QSR), good laboratory practice (GLP), good clinical practice (GCP), and EU regulatory requirements.

Her academic background includes a PhD/MPhil from University of London, St. George's Hospital Medical School, UK; Master of Science in biochemistry from the University of London, Chelsea College, UK; a Bachelor of Science in applied biology from Thames Polytechnic, UK; and an executive master's degree in business administration from Florida Atlantic University, School of Business.

Dr. Venky Gopalaswamy is a business improvement services group leader in a Fortune 100 company. In his role, he provides leadership to the corporation on business improvement methodologies such as Six Sigma, Design for Six Sigma (DFSS), Lean Management, and Change Management. In addition, he supports innovation and product development efforts across the corporation to enable successful and innovative product, service, and process development. Dr. Gopalaswamy has coached, mentored, and trained many company associates throughout their DFSS journey. His career spans safety-critical product/process technology development and quality/reliability engineering roles in medical device, high-tech, and automotive industries.

Dr. Gopalaswamy is a company-certified Six Sigma Black Belt and is currently pursuing Master Black Belt certification. He earned his PhD and MS degrees in industrial and systems engineering from the University of Illinois, and his undergraduate degree in mechanical engineering from India. He is also an ASQ-certified CQM, CRE, and CQE and is a senior member of ASQ and PDMA. He has published and presented many papers and has coauthored two books: *Practical Design Control Implementation for Medical Devices* and *From Design Control to Six Sigma for Medical Devices.*

chapter one

Introduction

We decided to write this book for one simple reason. The Food and Drug Administration's (FDA) requirements for combination products have evolved significantly in the last few years, and this key area of product growth is going to continue on a positive trend in the future with a major impact to industry and society.

Many medical device, pharmaceutical, and biologics companies are looking for practical guidelines on how to develop combination products successfully in spite of all the challenges resulting from merging of technologies and systems, and how to implement these products into the market successfully and in a timely manner. There is limited information available on this topic in a concise single book, and hence we felt there was a significant need in writing a book that looked practically at the challenges facing companies that are developing or planning to develop combination products.

Medical device, pharmaceutical, and biologics industries worldwide have grown tremendously over the past 30 years. Key milestones that have contributed to this growth include, but are not limited to, the following:

- Publication of good manufacturing practices (GMP) by the FDA in July 1978
- Development and marketing of recombinant products such as genetically engineered human insulin and hepatitis vaccine in the 1980s
- Mapping of the human genome in the 1990s

The development and marketing of combination products in this decade is another key milestone for these industries.

Achieving this milestone is significant since it signaled a key strategic shift for the U.S.-based pharmaceutical, medical device, and biologics industry, which is struggling with patent expirations, numerous litigations, competition from generics, rapid and explosive growth in China and India, and so forth. The combination product market is generally seen as an area with high growth potential and greater profit margins while providing significant positive benefits to patients. For example, pharmaceuticals facing patent expirations can get a new life if paired with a device as a combination product, thereby providing some relief to the industry.

Another reason why combination products as a product family are on the rise with their ever-increasing breakthrough novel technologies is that they may aid patient care in the long term, resulting in significant reduction in healthcare costs for society. Additionally, combination products may also provide numerous therapeutic advantages by both potentially increasing the effectiveness of product components and reducing systemic side effects by enabling local delivery.

Medical device, pharmaceutical, and biologic products have been long regulated in the United States as well as other countries. Many guidance documents exist to help in the design and development of these products. Although there is significant information around the guidances and FDA websites on development and testing of devices, drugs, and biological products, these are addressed as individually.

This approach has resulted in these industries gaining substantial and in-depth knowledge in regulations. It has also resulted in these industries having the ability to develop and launch medical devices, pharmaceuticals, or biologics individually with a fair amount of success in meeting the regulations. Regulators are specialized in inspecting these companies within their industry segment. These are laudable achievements, and the industry and the regulatory agencies must be commended for them.

However, the arrival of combination products shook this specialized approach to developing, marketing, and inspecting. For example, pharmaceutical and biologics companies developed products over a long time horizon (e.g., 10 years), while medical device companies were faced with relatively shorter development times (e.g., 3 to 5 years). When a medical device company is tasked to develop a combination drug-device product, it is only natural to expect the company's culture to assume that since a majority of medical device regulations will be applicable, the overall development time will be almost the same as that of a medical device. However, both the industry and the regulatory bodies now know that this expectation, along with other expectations on regulatory requirements, is simply not valid. In order to guide the industry better, the need for guidance documents that focus on the technical, scientific, regulatory, and quality issues when combining drug, device, and biological constituent parts to develop a combination product is just being addressed. This learning continues for both the industry and the regulatory bodies through joint industry-FDA meetings, professional society conferences, and so on. For its part, the Regulatory Affairs Professional Society (RAPS) in the United States dedicated separate tracks for combination products at its 2005 and 2006 annual conferences.

The correct regulatory pathway, product development systems, and implementation of quality systems for combination products, when combined together, pose unique challenges. When we throw into this mix the necessary people competencies needed to develop these products, one can easily see the complexity in developing and marketing combination products. Does that mean that companies give up their plans to grow by developing and launching combination products? We do not think so, since based on current industry experience and future projections, it appears that the rewards are significant for both the industry and society.

Our combined experience over the years in the pharmaceutical and medical device industries observing and dealing with product development, compliance, and regulatory challenges, including those for combination products, has given us insights into the practical aspects of some of the daily challenges that face product development teams. In this book, we focus on sharing our insights on the developmental processes for combination products, which may include products that result in a final product combination of devices, biologics, and drugs. It is our strong belief that this book will provide clarity to the confusion and challenges mentioned earlier in developing combination products.

In regards to combination products, the FDA has significant information in terms of guidance documents and websites on development and testing of devices, drugs, and biological products, which are addressed as individual products.

The guidances that would address the technical, scientific, regulatory, and quality issues when combining drug, device, and biological constituent parts to develop a combination product have just recently been addressed.

There are many publications that separately address the development of medical devices, pharmaceuticals, and biologics (refer to Center for Biologics Evaluation and Research [CBER], Center for Drug Evaluation and Research [CDER], and Center for Device and Radiological Health [CDRH]). However, the field of combination product development is so new that there are very few publications, including the FDA's guidance document. None of these publications go to the depth that will help medical device, pharmaceutical, and biologics industries in the development of combination products.

This book focuses on the developmental and regulatory processes for combination products, which may include products that result in a final product combination of devices, biologics, and drugs. This includes challenges, such as developmental, technical, regulatory, manufacturing, and launching of these novel combination products, that industry currently faces.

In many instances drug-eluting stents (DESs) will be used as an example to explain the routes of regulatory, quality, or developmental thinking in bringing these products to market since DES, as a product family, has the greatest amount of history and experience within the industry. The focus of this book is not on combinations that can be a product comprised of a cosmetic and pharmaceutical in nature.

By writing this book, we also want to fill the void in the availability of published, practical-experience-based reference material in the areas of combination products, and the technical and developmental challenges faced by industry in addressing these situations. We will also provide sufficient clarity on the design, development, validation, and control of the processes that are needed for manufacturing these products.

The following are the major features of this book:

- Industry's first book that provides an in-depth look at combination product development

- Fills a void to support successful development of one of the most significant product platforms in the healthcare industry in a long time
- Includes practical advice based on many years of experience in combination product development, quality, and regulatory areas
- Helps meet FDA and other regulatory body guidelines

In Chapter 2, we will help the reader gain a basic understanding of definitions of combination products. Chapter 3 focuses on ensuring successful development of combination products and the value these products bring to the patients. It also discusses the establishment of a design and development plan during early product development for successful development of combination products. Chapter 4 discusses meeting the expectations of FDA and other regulatory agencies and also focuses on the regulatory requirements for combination products in the EU, Canada, Japan, China, and India. The regulatory challenge for combination products that may result in a final product combination of devices, biologics, and drugs is significant.

Chapter 5 focuses on development and implementation of support systems for combination products, such as resources, IT, technical, facilities, quality, regulatory, clinical, and manufacturing. Chapter 6 focuses on key manufacturing challenges faced by organizations that have entered or are contemplating entering the combination product market. Chapter 7 focuses on how to address some of the challenges and pitfalls during the development of combination products, as well as how to avoid them. Chapter 8 discusses the postlaunch compliance requirements for companies developing and manufacturing combination products. Chapter 9 discusses agency audits and challenges and how to address these challenges, and Chapter 10 summarizes the content in the previous chapters and provides the reader with meaningful insights and conclusions. Finally, references are provided at the end of each chapter.

Market

The primary audience for this book is anyone responsible for developing and implementing a product or process to comply with the FDA's combination product guidelines. Specifically, it is targeted at individuals within the industry that are involved in combination products from a development, regulatory, clinical, quality, or manufacturing elements perspective.

This includes engineers in product and process design, development, and implementation in small, medium, and large medical device, pharmaceutical, and biologics companies in the United States, as well as those outside of the United States. This book can also be used by companies that have implemented or are in the process of implementing a quality system for combination products within their organization and are looking for "watch-outs" as they develop the systems.

The book also serves the needs of other product or process development team members, including, but not limited to, representatives from marketing, quality, regulatory compliance, and clinical that may benefit from the knowledge contained in this book.

International potential

This book will also help professionals and management at healthcare companies, both inside and outside of the United States, to develop combination products that can meet FDA regulations and other regulations, such as EU, Japan, and so forth. Content in this book is useful to global markets and industries developing and marketing combination products.

Companies that develop and market combination products successfully in the next decade will be the ones that reinvent the way people are diagnosed and treated for life-threatening diseases. Combination products will not only reward these companies financially but also enhance their reputation in society. We hope you will come away knowledgeable and excited by the possibilities of successfully developing and marketing combination products.

chapter two

Overview of combination products

Combination products are comprised of two or more regulated components, that is, drug-device, device-biologic, biologic-drug, or drug-device-biologic. These can be either physically or chemically combined as a single entity, or these can be two different products that may be packaged together or separately. If packaged together, the labeling of each product needs to dictate that they are to be used together in some way. Later in this chapter we will introduce the definitions used in the Food and Drug Administration's (FDA) guidelines.

Combination product classification

Combination products can be generally categorized into four separate classes based on their specific components: novel drug delivery systems, traditional drug delivery systems, drug-enhanced devices, and regenerative medicinal products.

Novel drug delivery systems

Examples in this category include patches, transdermal or intradermal injections, inhalation devices, sprays, and drug-eluting disks, which typically combine existing drugs with *new delivery devices*. These products are designed to improve the convenience and comfort of administration, and drug effectiveness through localized administration, and to enable delivery of a drug through a different administrative route other than oral, subcutaneous, or intramuscular injections. In many cases, these changes to administration or drug formulations and bioavailability could potentially increase the drug development technological challenges, although these products have a moderate complexity. These products would be primarily governed by the Center for Drug Evaluation and Research (CDER) through its regulatory pathway for drugs since the primary therapeutic mode of action would be drug related in these cases.

Traditional drug delivery systems

These products combine or package drugs together with injection devices to improve convenience of administration. This would include pen-based delivery systems, drug pumps, profiled syringes, and auto injectors. Since the drug-device interface is relatively simple in these products, the components in this instance can be developed separately and integrated in the later stages of the product development cycles. These components can also be regulated separately using the established regulatory regimes for drugs and devices.

Drug-enhanced devices

Products such as drug-eluting stents, bone cements with antimicrobial agents, coated catheters, other devices with antimicrobial coatings, and infective sutures can enhance the functionality, performance, or efficacy of these devices.

In large instances, these products combine existing devices with existing drugs. In these cases, the drug-device interface is often novel and is oftentimes critical to the performance of the combination product. As a result, development of these products is more complicated than similar device-only products. In these cases, since the primary therapeutic action stems from devices (CDRH), device regulators would primarily govern these products, with a secondary review from (CDER) drug-related regulatory agencies.

Regenerative medicinal products

These are products that combine devices with biologically active substances to facilitate/enhance healing and regeneration of damaged tissues. The device often serves as the support structure for the growth of the biologic component with the product often being an implant. Examples of such products include absorbable meshes for bone growth, Dermagraft (human fibroblast–derived dermal substitute), artificial replacement organs (such as the bioartificial pancreas), and coated spinal fusion cages with recombinant human bone morphogenic proteins. These products are among the most complex combination products, as they have to consider the interaction between the product and the body's physiological response to the product. As a result, the development process for such products is also extremely complicated and integrated because the components under development have to be tightly coupled. The lead review and oversight by the relevant agency would vary in these cases, as the primary mode of action for these combination products may vary on the specific combinations.

Examples of combination products

An example of a combination product is the tapered-metallic spinal-fusion cage, which is used in combination with recombinant human bone morphogenic protein, placed on a reabsorbable collagen sponge. The recombinant protein induces formation of new bone. The Food and Drug Administration's

(FDA) Center for Device and Radiological Health (CDRH) regulated this product via the premarket approval (PMA) process, in consultation with the Center for Biologics Evaluation and Research (CBER). In addition, several photoactivated drugs have been approved by the FDA. These drugs, which are regulated by the CDER, require a specially designed light source to activate the drug. The photoactivation light sources are cross-labeled with the drug and regulated by CDRH via the PMA path.

Products that fall under the umbrella of combination product include, but are not limited to, the following:

1. *Drug-eluting stents* for cardiovascular or endovascular applications that may prevent restonosis (e.g., Cypher, Taxus). The drug-coated coronary artery stents are an example of a combination product that includes a device, the metallic stent with a variety of drugs coated on its surface. The drugs are intended to enhance the physical effect of the device by preventing closure of the vessel after treatment. This product is an example of a medical device that incorporates a drug as an ancillary medicinal substance. While many products launched prior to drug-eluting stents could potentially be classified as combination products, these were the first ones to be regulated as such. In this case, the product's primary mode of action (PMOA) is affected by the medical device's mechanical function (the stent's struts hold the artery open), with the drug present to prevent excessive scar tissue from forming on the stent and potentially causing a new obstruction. Here, the drug's action is secondary to the device. In this instance, the FDA concluded that the primary mode of action for the combination product is that of the device component and assigned CDRH primary responsibility for premarket review and regulation. We will present a detailed discussion on PMOA later in this chapter.
2. *In vitro diagnostics* used to determine whether patients are eligible for treatment with a specific drug.
3. *Novel orthopedic implants* may facilitate the regeneration of bone required to permanently stabilize these implants by incorporation of additional proteins (e.g., infuse bone graft system-lyophilized bone morphogenic protein [BMP], collagen, and spinal cage).
4. *Iontophoretic transdermal systems* that provide on-demand systemic delivery of fentanyl or other potent opiods for analgesia or sedation.
5. *Drug-device inhalation systems* may provide a novel means of delivering insulin for diabetics that may decrease or reduce the need for injections.
6. *Diagnostic devices based on genomic technology* may assist in determining the suitability of certain types of products or alert to adverse risks.
7. *Development of Gliadel,* a controlled-release drug product for the treatment of malignant glioblastomas; the material comprising the implanted wafer was specifically synthesized for local controlled release of the chemotherapy agent. This highly toxic drug could thus be delivered locally, only at places where it is needed, thus minimizing systemic toxicity.

Biologic-device combination product

By definition, this contains a biological component and a device component, where both components function in a close manner. Current advancements in technology such as tissue engineering have created an interesting and challenging environment in respect to the regulatory pathway for approvals as well as the FDA branch having primary jurisdiction over the combination product. Novel technologies and products fall in an uncertain area of which branch, that is, CDER, CBER, or CDRH, would have the main say on the regulatory pathway for a specific product. An example of biologic-device combinations can be seen with Bioartificial organs. Biological components of a bioartificial organ could include either isolated cells, cell lines, or living tissues, and synthetic components may include biopolymers, plastics, or permeability-selective membranes, among other parts. An example of a biologic-device combination product is the HepatAssist system.

The HepatAssist Liver Support System (LAS) is considered a bioartificial organ (liver), designed to temporarily support patients with severe liver failure. The biological component is uniquely processed, cryopreserved porcine hepatocytes, and its medical device component is a unique, hollow-fiber cartridge. LAS was designed to support patients with acute liver failure until either liver regeneration takes place or a liver allograft is available for transplantation. As a combination product, LAS could have been designated as a device or a biologic. The most critical component of the combination product in this case was determined to be biological, and as a result, in 1994, the CBER was the main regulatory agency with consultation with CDRH.

Bioartificial organs (tissue engineered)

Products containing both biologic (cells) and device (matrix) regulatory authority rest with CBER with CDRH consult. Cell- or tissue-based combination products are complex, requiring input across FDA; there are guidances available, but numerous concerns regarding source tissue, contaminants, reagents, viability, and compatibility still exist.

Recently approved combination products

Table 2.1 lists more than twenty recently approved combination products in chronological order, and Table 2.2 provides a summary of the list from Table 2.1. From the summary it can be seen that a majority of the approvals came from CDRH where the PMOA is a device. We will now look at some of the approved products to provide additional insight to the reader on combination products and the definitions associated with them.

Table 2.1 Recent Examples of Combination Product Approvals by FDA (As of March 2007)

Product approved	Marketed by/ manufactured by	FDA lead center	Date approved	PMOA
Absorbable collagen sponge with genetically engineered human protein	Medtronic Sofamor Danek, Inc.	CDRH	March 9, 2007	Device
Device-biological product gel for surgical hemostasis	Orthovita, Inc.	CDRH	June 16, 2006	Device
Iontophoretic transdermal system for fentanyl	Ortho-McNeil, Inc./ ALZA Corporation	CDER	May 19, 2006	Drug
Transdermal patch for attention deficit hyperactivity disorder	Shire US, Inc./Noven Pharmaceuticals, Inc.	CDER	April 10, 2006	Drug
Transdermal patch for depression	Bristol-Myers Squibb/Somerset Pharmaceuticals, Inc.	CDER	February 28, 2006	Drug
Inhaled insulin combination product for diabetes	Pfizer, Inc.	CDER	January 27, 2006	Drug
Surgical mesh with antibiotic coating	American Medical Systems, Inc.	CDRH	December 14, 2004	Device
Dermal iontophoresis system	Empi, Inc./N ovocol Pharmaceutical of Canada, Inc.	CDER	October 26, 2004	Drug
Iontophoretic drug delivery patch and controller	Vyteris, Inc.	CDER	May 6, 2004, and August 20, 2003	Drug
Methylaminolevulinate cream with halogen light source	PhotoCure ASA	CDRH	July 8, 2004	Device
Absorbable collagen sponge with genetically engineered human protein	Wyeth Pharmaceuticals	CDRH	April 30, 2004	Device
Spinal fusion putty	Stryker Biotech	CDRH	April 7, 2004	Device

(Continued)

Table 2.1 *(Continued)*

Product approved	Marketed by/ manufactured by	FDA lead center	Date approved	PMOA
Paclitaxel-eluting coronary stent system	Boston Scientific Corporation	CDRH	March 4, 2004	Device
Influenza virus vaccine, live, intranasal	Medimmune Vaccines, Inc.	CBER	June 17, 2003	Biologics
Tositumomab and iodine I 131 tositumomab	Corixa Corporation	CBER	September 25, 2003	Biologics
Demagraft human fibroblast–derived dermal substitute	Smith and Nephew Wound Management	CDRH	July 7, 2003	Device
Glucose monitor/ insulin pump	Medtronic MiniMed, Inc./Becton Dickinson	CDRH	July 7, 2003	Device
Antibiotic bone cement	Howmedica Osteonics Corporation	CDRH	May 6, 2003	Device
Sirolimus-eluting coronary stent	Cordis Corporation	CDRH	April 24, 2003	Device
Fibrin sealant	OMRIX Biopharmaceuticals, Ltd.	CBER	March 21, 2003	Biologics
Dermal collagen implants for aesthetic use	Inamed Corporation	CDRH	March 11, 2003	Device
Peginterferon alfa-2a in combination with ribavirin	Hoffman-La Roche, Inc.	CBER	December 3, 2002	Biologics
Lumbar tapered fusion device with genetically engineered human protein	Medtronic Sofamor Danek, Inc.	CDRH	July 2, 2002	Device

Source: http://www.fda.gov/oc/combination/approvals.html.

Breath test combination products

Since 1992, the FDA has received numerous Requests for Designation (RFDs) for combination product diagnostic breath tests where the drug component is an isotope-labeled (typically ^{13}C) substrate and the device components capture or analyze exhaled breath for detection of labeled carbon dioxide or other gases. These breath test combination products are used for the diagnosis of *Helicobacter pylori*, gastric emptying disorders, carbohydrate malabsorption,

Table 2.2 Summary of Recent Examples of Combination Product Approvals by FDA

	FDA Lead Center			
PMOA	CBER	CDER	CDRH	Percent
Biologics	4			17%
Device			13	57%
Drug		6		26%

intestinal bacterial overgrowth, and insulin resistance, liver function, for monitoring enzyme activity, for assessment of small intestinal function, and for use in pharmacological research. In some cases, such as diagnosis of *H. pylori*, the presence, absence, or rate of release of isotope-labeled carbon dioxide in exhaled breath is intended to be indicative of the presence or absence of the disease or condition in question. In these cases the determination by the FDA was that the PMOA of such combination products was attributable to the device component's role in the *in vitro* diagnosis of the disease or condition in question, while the drug component plays a secondary role in acting as the diagnostic substrate. The FDA assigned the Center for Devices and Radiological Health to be the lead agency center for reviewing these products. The FDA also determined that two marketing applications were not necessary for a diagnostic breath test combination product. In this recent case, the FDA determined that the PMA provisions of the act (21 CFR 814) would enable the agency to determine the safety and effectiveness of both the device and drug components of the combination product. CDRH will consult or collaborate with CDER as appropriate on issues such as chemistry and manufacturing, pharmacology and toxicology, and clinical issues related to the drug component.

Drug-device combination catheter lock/flush solutions

The FDA has received RFDs for catheter lock/flush solutions containing an anticoagulant or an antimicrobial in a solution (e.g., water or saline solution). These catheter lock/flush solutions are intended to be used to maintain catheter patency. The RFDs requested clarification of the FDA center with primary jurisdiction over these products, as well as the statutory provisions under which these products would be reviewed and regulated.

In these cases, the FDA determined that the solution component of these products (e.g., water or saline solution) acts by physically occupying space within the catheter and exerting pressure on the patient's circulating blood. In this way, the patient's blood is prevented from backfilling into the catheter and clotting. The FDA concluded that in acting in this manner, the solution component of the product meets the definition of a device in that it affects the structure or function of the body, and does not achieve its primary intended purposes through chemical or metabolic action.

The FDA also determined that in these cases, the anticoagulant or antimicrobial component of the catheter lock/flush solutions was intended to affect the structure and function of the body of humans by acting chemically on microorganisms and preventing thrombotic occlusions with the catheter. The FDA concluded that such components meet the definition of a drug in that they are intended to affect the structure or function of the body of humans.

For these reasons, the FDA concluded that the catheter lock/flush solutions described in these RFDs consisting of a solution (e.g., water or saline solution) and an anticoagulant or antibiotic are combination products within the meaning of section 503(g) of the federal *Food, Drug, and Cosmetic Act* (21 USC § 353(g)). Accordingly, review responsibility for these products was assigned based on the agency's determination of the products' primary mode of action (section 503(g)(1) of the act, 21 USC § 353(g)(1), and 21 CFR § 3.4).

The FDA determined that the PMOA of these catheter lock/flush solutions in maintaining catheter patency was attributable to the device component's role in occupying space and applying pressure within the catheter. The FDA also concluded that the drug component of the combination product plays a secondary role by, for example, chemically acting on microorganisms or preventing thrombotic occlusions within the catheter.

Biologic-device: Vitagel Surgical Hemostat

The Vitagel Surgical Hemostat is a product that assists the body in clotting blood. Vitagel is absorbed by the body after performing its function. Vitagel includes the biological component thrombin, an enzyme that assists in the clotting of blood. Vitagel is intended to assist in clotting when conventional means fail or are impractical. The primary mode of clotting for Vitagel is the formation of a protein (collagen/fibrin) clot that serves as a physical barrier to blood flow. The thrombin component helps to form a clot by converting the protein fibrinogen to fibrin during clotting. Vitagel is used during surgical procedures (except neurosurgery and eye surgery) as an adjunct to clotting when control of bleeding using suture or other conventional procedures is not effective or seems impractical. The product was shown to be effective in controlling bleeding in hepatic, general, cardiac, and orthopedic surgical procedures.

Differently regulated constituent parts (drug, device, biological product)

Physically or Chemically Combined into One Entity (21 CFR 3.2(e)(1))
Examples include the following physically, chemically, and otherwise single entities:

- Biologic monoclonal antibody with a therapeutic drug
- Device coated, impregnated with a drug or biologic
- Drug-eluting stent, pacing lead steroid-coated tip
- Skin substitutes with cellular components

- Orthopedic implant with growth factors
- Prefilled drug or biologic product delivery device
- Syringes, insulin injector pens, metered dose inhalers, transdermal patches

Copackaged (21 CFR 3.2 (e)(2))

- Drug/biologic product packaged with a delivery device, or packaged with a diagnostic test.
- Separately marketed, both are required, and when mutually conforming, labeling is needed.
- Photodynamic therapy drug and laser/light source.
- Drug requiring specific device for administration.
- Diagnostic device required for use of a specific drug or biological product.

Other parts include *separately provided cross-labeled products* (21 CFR 3.2(e)(3)) and *separate investigational products with proposed cross-labeling* (21 CFR 3.2(e)(4)).

GMP guidance

For a combination product, the current good manufacturing practices (cGMP) and quality system regulations (QSR) are similar; however, the quality system needs to be tailored to the product for which they were designed. The combination product has to be in compliance with the regulations for the constituent part. Each constituent part is subject to its governing GMP regulations before combination. During and after combination (21 CFR 3.2(e)(1) or (e)(2)), both sets of regulations would apply. Compliance with both regulations can generally be achieved using either (e.g., by using the system in place at a facility); however, parallel systems are not needed.

Example 1: For a Combination Product That Is a Device-Drug

> The device manufacturer already has a quality system or QSR in place at its facility. The operating manufacturing control system at this site for the drug constituent would include adding the drug regulations indicated in Table 2.3.

Example 2: For a Combination Product That Is a Drug-Device

> The drug manufacturer would already have a cGMP system in place (Parts 210 and 211) at its facility. The operating manufacturing control system at this site for the device constituent would include the quality system regulations as indicated in Table 2.4.

Table 2.3 Drug Provisions to Consider

211.84	Testing and approval or rejection of components, drug product containers, and closures
211.103	Calculation of yield
211.137	Expiration dating
211.165	Testing and release for distribution
211.166	Stability testing
211.167	Special testing requirements
211.170	Reserve samples

Table 2.4 Medical Device Provisions to Consider

820.30	Design controls
820.50	Purchasing controls
820.100	Corrective and preventive action

A list of definitions and terminology used in conjunction with combination products and their regulation is provided below:

Combination product: A product comprised of any combination of a drug and a device, a biological product and a device, a drug and a biological product, or a drug, device, and a biological product.

As defined in 21 CFR Part 3, the term *combination products* includes the following:

1. A product comprised of two or more regulated components; i.e., drug/device, biologic/device, drug/biologic, or drug/device/biologic, that are physically, chemically, or otherwise combined or mixed and produced as a single entity.
2. Two or more separate products packaged together in a single package or as a unit and comprised of drug and device products, device and biological products, or biological and drug products.
3. A drug, device, or biological product packaged separately that according to its investigational plan or proposed labeling is intended for use only with an approved individually specified drug, device, or biological product where both are required to achieve the intended use, indication, or effect and where upon approval of the proposed product the labeling of the approved product would need to be changed; e.g., to reflect a change in intended use, dosage form, strength, route of administration, or significant change in dose.

4. Any investigational drug, device, or biological product packaged separately that according to its proposed labeling is for use only with another individually specified investigational drug, device, or biological product where both are required to achieve the intended use, indication, or effect.
   - A constituent part of a combination product is an article in a combination product that can be distinguished by its regulatory identity as a drug, device, or biological product, as defined in section 21 USC 321, federal *Food, Drug, and Cosmetic Act*, and 42 USC 252(i), Public Health Service Act, and as set forth in 21 CFR 3.2(k).
     For example, a device coated or impregnated with a drug has two constituent parts: the device constituent and the drug constituent. A combination product is composed of two or more constituent parts.
5. *Mode of action*: "The means by which a product achieves its intended therapeutic effect or action. For purposes of this definition, 'therapeutic' action or effect includes any effect or action of the combination product intended to diagnose, cure, mitigate, treat, or prevent disease, or affect the structure or any function of the body."
   - Products may have a drug, biological product, or device mode of action. Because combination products are comprised of more than one type of regulated article (biological product, device, or drug), and each constituent part contributes a biological product, device, or drug mode of action, combination products will typically have more than one mode of action.
6. *Primary mode of action (PMOA)*: "The single mode of action of a combination product that provides the most important therapeutic action of the combination product." The most important therapeutic action is the mode of action expected to make the greatest contribution to the overall intended therapeutic effects of the combination product.
   - For example, if the PMOA of a device-biologic combination product were attributable to its biological product constituent, the agency component responsible for premarket review of that biological product would have primary jurisdiction for the combination product.
7. *A constituent part of a combination product*: An article in a combination product that can be distinguished by its regulatory identity as a drug, device, or biological product, as defined in section 201 of the *Food, Drug, and Cosmetic Act* or 351(i) of the Public Health Service Act.
8. *Drugs*:
   - Articles recognized in the official United States Pharmacopeia or National Formulary, or any supplement to either of these:
     a. Intended for use in the diagnosis, cure, mitigation, treatment, or prevention of disease in humans or other animals.
     b. Articles (other than food) intended to affect the structure or any function of the body of humans or other animals.

c. Substance whose primary intended use purpose is achieved through chemical action or by being metabolized by the body.

9. *Adverse drug experience*: Any adverse event associated with the use of a drug in humans, whether or not considered drug related, including the following:
   - An adverse event occurring in the course of the use of a drug product in a professional practice
   - An adverse event occurring from a drug overdose whether accidental or intentional
   - An adverse event occurring from drug abuse
   - An adverse event occurring from drug withdrawal and any failure of pharmacological action
10. *Disability*: A substantial disruption of a person's ability to conduct normal life functions.
11. *Life-threatening adverse drug experience*: Any adverse event drug experience that places the patient, in view of the initial reporter, at immediate risk of death from the adverse drug experience as it occurred; this does not include an adverse drug experience that, had it occurred in a more severe form, might have caused death.
12. *Serious adverse drug experience*: Any adverse drug experience occurring at any dose that results in any of the following outcomes:
   - Death, a life-threatening adverse drug experience, inpatient hospitalization or prolongation of existing hospitalization, a persistent or significant disability/incapacity, or a congenial anomaly/birth defect
13. *Unexpected adverse drug experience*: Any adverse drug experience that is not listed in the current labeling for the drug product. This includes events that may be symptomatically and pathophysiologically related to an event listed in labeling, but differ from the event because of greater severity or specificity.
14. *Devices*:
   - An instrument, apparatus, implement, machine, contrivance, implant, *in vitro* agent, or similar or related article, including any component, part, or accessory, that is:
     a. Recognized in the official United States Pharmacopeia or any supplement.
     b. Intended for use in the diagnosis of disease or other conditions, or in the cure, mitigation, treatment, or prevention of disease, in humans or other animals, or metabolized for the achievement of its primary intended purposes.
     c. Intended to affect the structure or any function of the body of humans or other animals, and which does not achieve its primary intended purposes through chemical action within or on the body of humans or other animals, and which is not dependent upon being metabolized for the achievement of its intended purposes.

15. The FDA classifies medical devices into one of three categories:
    - *Class I* is for low-risk devices such as manual surgical instruments, medical disposable scissors, dental flosses, and so on.
    - *Class II* is for devices that can not be classified in class I because there is insufficient information to show that the general controls alone are adequate to ensure their safety and effectiveness.
    - *Class III* is for devices that are used to support or sustain human life, present an unreasonable risk of injury, or can not be classified in class I or II because there is insufficient information to determine that class I or II controls provide reasonable assurance of their safety and effectiveness.

    The various classes identify the amount of regulation a device will encounter.
16. *Medical device reporting (MDR)*: The mechanism for the Food and Drug Administration to receive significant medical device adverse events from manufacturers, importers, and user facilities, so they can be detected and corrected quickly. If you are a consumer or health professional, you should use the MedWatch program for reporting significant adverse events or product problems with medical products.
17. *Biologics* (section 351, Public Health Service Act): Any virus, therapeutic serum, toxin, antitoxin, vaccine, blood, blood component or derivative, allergenic product, or analogous product applicable to the prevention, treatment, or cure of diseases or injuries of humans.
18. *Not combination products*: The following products are not classified as combination products:
    - Drug-drug, device-device, or biologic-biologic combinations
    - General drug or biologic delivery devices (e.g., unfilled syringe or infusion pump) not intended for use with an individually specified drug or biologic product
19. *OCP*: Office of Combination Products at the Food and Drug Administration. The OCP was established on December 24, 2002, as required by the Medical Device User Fee and Modernization Act of 2002 (MDUFMA). The law gives the office broad responsibilities covering the regulatory life cycle of drug-device, drug-biologic, and device-biologic combination products. However, the primary regulatory responsibilities for, and oversight of, specific combination products will remain in one of three product centers—the Center for Drug Evaluation and Research, the Center for Biologics Evaluation and Research, or the Center for Devices and Radiological Health—to which they are assigned.
20. *CDER*: The Center for Drug Evaluation and Research (CDER) at the Food and Drug Administration; regulates drug products. CDER ensures that prescription and over-the-counter drugs, both brand name and generic, work correctly and that the health benefits outweigh known risks.
21. *CBER*: The Center for Biologics Evaluation and Research (CBER) at the Food and Drug Administration; regulates biological products. Current

authority for this responsibility resides in section 351 of the Public Health Service Act and in specific sections of the *Food, Drug, and Cosmetic Act.*

22. *CDRH*: The Center for Devices and Radiological Health (CDRH) at the Food and Drug Administration; responsible for ensuring the safety and effectiveness of medical devices and eliminating unnecessary human exposure to man-made radiation from medical, occupational, and consumer products.
23. *RFD*: A Request for Designation; the mechanism by which sponsors can request a formal determination of either the regulatory identity of a product as a drug, device, biological product, or combination product, or which center will have primary jurisdiction for premarket review and regulation of a combination product.
24. *cGMP*: Current good manufacturing practice; refers to the current good manufacturing practice regulations for drugs and most biological products under 21 CFR Parts 210 and 211, for certain biological products under 21 CFR Parts 600–680, and the quality system regulations for devices under 21 CFR Part 820.
25. *Manufacture*: Refers to the methods to be used in, and the facilities and controls to be used for, the manufacture, processing, packing, or holding of a drug (21 CFR 210.01(a)), and those used for the design, manufacture, packaging, labeling, storage, installation, and servicing of all finished devices intended for human use (21 CFR 820.1(a)). Additionally, the term can also refer to the methods and facilities for certain biological products that are considered to supplement, not supersede, the drug provisions, unless the regulations explicitly provide otherwise (21 CFR 210.2(a)).
26. *Manufacturer*: Refers to any person who would be required to comply with current good manufacturing practice regulatory requirements for drugs, biological products, devices, or combination products.

European union (EU) definitions

1. *Competent authority (CA):* A body with authority to act on behalf of the government of the member state to ensure that the requirements of the Medical Devices Directive are transposed into national law and applied. The government of each member state is required to appoint a CA responsible for medical devices. The CA reports to the minister of health in the member state. The CA in one member state does not have jurisdiction in any other member state, but he or she does exchange information and try to reach common positions. The CA is responsible for the enforcement of appointing notified bodies, supervising the work of the notified bodies, surveillance of medical devices on sale in his or her own member state, evaluation of adverse incidents, approval to start clinical investigations, and registering class I devices and in-vitro diagnostics (IVDs) on sale in

his or her own member state. A CA is not obliged to appoint a notified body. However, he or she can appoint as many as apply that meet the criteria, and there is no limit. A few small countries do not have any notified bodies.

2. *Notified body (NB):* A certification organization (e.g., independent testing house, laboratory, or product certifier) authorized by the relevant member state's competent authority to perform the conformity assessment tasks specified in the Medical Devices Directive.
3. *MDD (93/42/EEC):* The European Union (EU) Medical Devices Directive; covers, among others, normal bandages, sports tape, scalpels, surface electrodes, infusion pumps, cardiac catheters, bone cement with antibiotics, intraocular lenses, wheelchairs, crutches, external pacemakers, and installations and apparatus in operating rooms. This directive, mandatory since June 14, 1998, requires non-European medical device manufacturers to appoint a European-authorized representative to sell their products in Europe.
4. *MPD (65/65/EEC):* The European Union's Medicinal Products Directive; covers "any substance or combination of substances presented for treating or preventing disease in human beings or animals," and "any substance or combination of substances which may be administered to human beings or animals with a view to making a medical diagnosis or to restoring, correcting or modifying physiological functions in human beings or in animals is likewise considered a medicinal product."

The definitions and abbreviations provided in this chapter should help the reader in reaching a common understanding of the terms typically used in the combination product industry, especially in the regulatory context. In the next chapter we will discuss approaches to ensure successful combination product development.

Bibliography

21 CFR, Part 58. *Good laboratory practice regulation,* 1987.

21 CFR, Part 210. *Current good manufacturing practice in manufacturing, processing, packing or holding of drugs,* 2006.

21 CFR, Part 210, *211. Drug cGMP regulation,* 2006.

21 CFR, Part 211. *General and current good manufacturing practice for finished pharmaceuticals,* 2006.

21 CFR, Part 314. *Applications for FDA approval to market of a new drug,* 2006.

21 CFR, Part 314. *Postmarketing reporting of adverse drug experiences,* 2006.

21 CFR, Part 600. *Biological products – general, 2001.*

21 CFR, Part 600. *Postmarketing reporting of adverse experiences,* 2006.

21 CFR, Part 610. *General biological products standards,* 2001.

21 CFR, Part 803. *Medical devices,* 2006.

21 CFR, Part 820. *Quality system regulations,* 2006.

Definition of primary mode of action of a combination product. Proposed rule, FDA. *Federal Register* 69(89):25527–25533, 2004.

Early development considerations for innovative combination products. Guidance for industry and FDA staff, Office of Combination Products, 2006.

Final rule on the definition of primary mode of action of a combination product. *Federal Register* 49848, August 25, 2005.

Guidance for industry and FDA current good manufacturing practice for combination products. Draft guidance, Office of Combination Products, September 2004.

Innovative systems for delivery of drugs and biologics: scientific, clinical and regulatory challenges. FDA workshop. Bethesda, Maryland, July 8, 2003.

Interface with other directives - medical devices/medicinal products. European Commission, 2001. http://ec.europa.eu/enterprise/medical_devices/meddev/2_1307-2001.pdf

Kramer, Mark D. Combination products: challenges and progress. *Regulatory Affairs Focus,* pp. 30–35, August, 2005.

chapter three

Ensuring successful combination product development

Benefits of developing combination products

In Chapter 2, we provided definitions, explanations, and examples to help the reader understand combination products. In this chapter we will provide the reader the reason to believe that combination products can be efficiently and effectively developed and launched. Though the challenges such as regulatory, quality, product development, clinical, and so on, are enormous with combination products, the benefits are also very significant.

From the patient or the healthcare provider perspective, products that fall under the combination product family are innovative and utilize cutting-edge technologies that can potentially improve patient care in terms of advanced, increasingly effective, convenient, and safer treatments for patients that could potentially result in less invasive modes of actions, shorter recovery times, and lower costs for the patient over time.

From the healthcare companies' perspective, these combination products can significantly increase patient care, revenues, and profits, and establish competitive advantage.

From the industry perspective, combination products provide a unique opportunity to grow the business substantially. Figure 3.1 provides insights into the global market size for medical devices in the year 2010. These numbers are based on numerous publications in the field of medical devices. This clearly indicates that companies that are first to market medical devices, including combination products, can reap significant revenue and profit growth benefits. For example, Cordis Corporation was able to reap the benefits of being first to market with its drug-coated stent. These drug-coated stents are tubes designed to be inserted into narrowed coronary arteries to help them remain open after balloon angioplasty. These stents then allow the normal flow of blood and oxygen to the heart. In the United States, with proper planning and relevant actions, the insurance providers can reimburse these products, as they can show significant healthcare economic benefits.

For device and drug companies, there are often common interests that make these partnerships invaluable. Both industries are extremely competitive to gain market shares for their respective products, and hence share

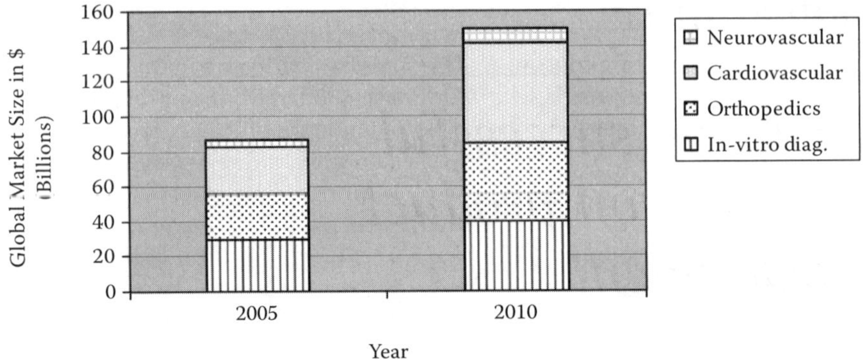

Figure 3.1 Global market size for medical devices.

a similar strategy for the promotion of their respective products. For the industry, developing combination products can also provide additional intellectual property protection. For example, if a pharmaceutical is coming off patent protection, the original owner of the patent can extend the protection if he or she can show unique uses/claims for the pharmaceutical as a combination product if it is delivered using a medical device, and lessen the competition it would face with the generics competitors once the drug patent expired. Real examples of financial benefits for companies can be seen with the recent development of certain combination products in the last few years, such as drug-eluting stent sales, with a billion dollars in sales annually.

While the benefits to patient and companies are significant, there are also significant risks associated with designing and developing combination products. These risks include, but are not limited to, poor business case, improper planning, inadequate resources, and clinical/technical challenges. It is our intent to shed as much light as possible on ways to mitigate these risks.

Product development in any industry is an extremely difficult challenge. It takes on an additional dimension in healthcare-related industries such as medical devices, diagnostics, pharmaceuticals, and biologics industries. (We will address these industries collectively as *healthcare product industries*.)

From discovery to commercial launch, product development efforts in these healthcare product industries consume enormous resources—time, people, and so on—to improve the probability of success of the product meeting the wants and needs not only of the regulators, but customers as well. Well-defined product development processes with integrated Six Sigma methodologies such as Design for Six Sigma (DFSS) help to ensure compliance with regulations and meet customer needs. When these industries want to develop combination products, the product development process and associated resources may be extended exponentially due to increased unknowns. For example, if a medical device company that is used to developing class II products (under FDA classification) desires to develop a combination product that includes a biologic agent, it has to know precisely how to develop and launch these products, which can be quite different from developing and launching class II medical devices alone. Unless there is a clear understanding and appreciation of the magnitude of unknowns, companies developing combination products can find themselves amidst disasters—financial, regulatory, and so on. On the other hand, if managed appropriately at all levels, development and launch of combination products can result in a very high return on investment.

Our aim in this chapter is to provide the reader with detailed combination product development approaches that can be applied regardless of his or her level of experience in product development in these regulated healthcare product industries.

Almost all companies developing products to diagnose or treat patient conditions have the ability to perform research. This research could be in the form of identifying new drug molecules, new medical device concepts, and so on. Competent people are hired and adequate infrastructure and funding are provided to support research. However, once the leadership of one or more of these companies decides to design, develop, and launch combination products to grow the business, it is important to provide adequate and competent resources, time, infrastructure, and funding to support these activities over and above what the company is used to.

Since most of these companies develop either medical devices or diagnostics or drugs or biologics, they usually lack resources, time, and know-how in developing combination products on their own. As a result, these companies typically resort to significant collaborations between drug and device, device and biologics, or drug and biologic companies. We have observed this on many occasions over the past few years, which have resulted in development, manufacture, and launch of such combination products. But can such collaborations address and provide the answer to companies not having their own system address combination product development? We do not think so, since appropriate elements of combination product must be included in these companies' own combination product development system. If not, these companies will be working hard to ensure linkages between two systems (e.g., a device and a drug development system) almost on a daily basis, which is detrimental to achieving business results. We have observed that many of the companies that either have launched combination products or have such products in their development pipeline have some sort of a hybrid product development road map with elements of the combination products (e.g., drug and device) brought together as one system.

Before we present our version of this hybrid product development road map for combination products, let us look at how products are typically developed in a traditional medical device, pharmaceutical, or biologics industry. Due to the implementation of the FDA's design control guidelines, most of the medical device and diagnostics manufacturers have implemented a design and development process that usually includes four or five stage/toll/phase gates and incorporates design control guidelines. See Figure 3.2 for a typical product development process in the medical device/diagnostics industry.

As you can see, the first stage focuses on developing a feasible concept internally or externally to the business. This feasible concept, once approved by the governance body at the stage review, then proceeds through the design stage to be further developed into a meaningful design using customer input, design technology such as finite-element modeling, and so on. Once this design is approved, the product development team starts focusing on developing a manufacturing process for the product using manufacturing technology that is unknown or known to the team. The process is then ready to be scaled up after proper validation/design transfer steps to prepare the product for launch/clinical trials. For a detailed review of medical device product development using Six Sigma techniques such as Design for Six Sigma (DFSS), we encourage readers to look into one of the coauthor's (Dr. Venky Gopalaswamy) books through CRC press, listed in the bibliography.

Compared to the medical device/diagnostics industries, a top-level review of the product development process in the pharmaceutical/biologics industries might reveal that the product development process with stages, gates, and so on, appears to be similar, but it must be emphasized that it is clearly not the same. Figure 3.3 depicts a typical product development process in pharmaceutical/biologics industries. During discovery, biologists and medicinal or protein chemists work together to research, create, evaluate, and recommend new drug/biologic entities that will satisfy unmet medical needs. Significant expertise and experience in a particular therapeutic area of interest is an absolute requirement, as is a thorough understanding of the overall drug discovery process. Critical to success at this stage is to have sufficient resources in required areas to build a well-rounded discovery team. Safety studies are performed during discovery to evaluate toxic effects of the drug/biologic entities on both animal and humans, leading to increased understanding of safety risks and increased ability to predict toxic effects.

Figure 3.2 Typical product development process in medical device/diagnostics industries.

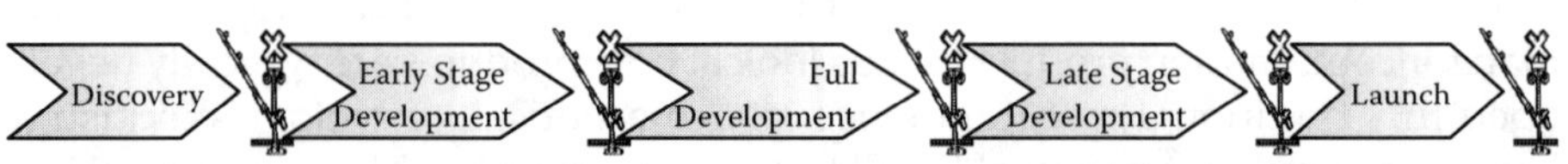

Figure 3.3 Typical product development process in pharmaceutical/biologics industries.

The road map to achieve control of design

The implementation of the FDA's guidelines, such as design control for medical devices and quality by design for pharmaceuticals, has resulted in manufacturers of medical devices, pharmaceuticals, and so forth, implementing a compliant and formal design and development process. As we presented in the earlier paragraphs, this process usually includes four or five stages/phases along with appropriate stage/phase/toll gates to govern that process.

According to Professor Nam Suh of MIT, the new product development process takes the design team through four different domains: customer, function, design, and process. In one of the author's earlier publications, *Six Sigma for Medical Device Design*, a fifth domain was introduced: innovation. This domain is present in combination product or, for that matter, all healthcare product industries because of the need for these companies to innovate constantly and develop meaningful product ideas prior to entering the customer domain. In addition to this, unlike many other industries, a large portion of product ideas in these industries comes from external sources such as end users, universities, and so on. These ideas, as well as those generated internally, must be evaluated and acted upon for many reasons, including but not limited to improving the company's intellectual property estate, assessing the market potential for the ideas, and assessing the early safety profile.

We will expand on this five-domain road map by formally adding one more domain that is critical for combination products and other healthcare product industries. This sixth domain will be called postmarket and is shown in Figure 3.4. This domain has always existed in these industries, but it was treated as an operations domain or marketing domain, or both. It was rarely viewed as a domain that needs to be considered during product development. The main difference between the postmarket domain and other five domains is that activities in this domain take place after the product is launched, but

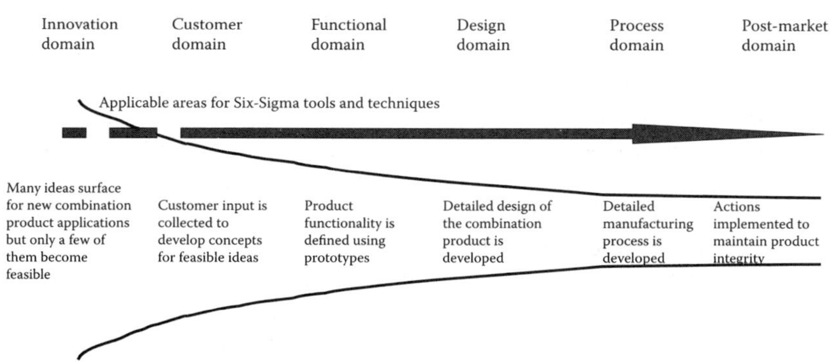

Figure 3.4 Combination product design and development domains.

Table 3.1 Combination Product Road Map Variations

	Medical device/diagnostics	Pharmaceuticals	Biologics
Medical device/diagnostics	Design control	Combination DP	Combination DB
Pharmaceuticals		Quality by design	Combination PB
Biologics			Quality by design

lots of planning and design efforts take place during product development to address these activities. We will explain these in detail later in this chapter.

This six-domain road map will be utilized to present our ideas on how to develop combination products successfully. We will present what activities typically happen in these domains to develop and launch combination products. It must be mentioned that this six-domain road map will have variations depending on the type of combination product that is developed. Table 3.1 highlights one such variation where the combination is presented based on applicable FDA regulations after the innovation domain. For example, if the combination product that is being developed is a medical device with a pharmaceutical (e.g., a drug-eluting stent), the appropriate road map is combination device-pharmaceutical (DP), where the road map includes the elements of the FDA's medical device design control, quality by design, and other applicable guidelines to develop the product. This table also illustrates the need to have an overall plan for design and development efforts.

Establishing a development plan

There are many items that need to be considered at the beginning when developing combination products. These include resource requirements, types of interactions needed with regulatory agencies, timing of these interactions, and so forth. Proper inclusion of these in a formal/documented design plan is essential for combination product development. This plan will help in speeding the development process since it can clearly describe the design and development activities and associated responsibilities for implementation. The plan shall identify collaborative efforts or interfaces needed to provide input to the design and development process. This, in turn, will highlight the needed skills in devices or drugs or biologics, depending on the combination product being developed.

The Food and Drug Administration released a guidance document in September 2006 titled *Early Development Considerations for Innovative Combination Products*. In that document certain key considerations are highlighted for healthcare product manufacturers. For example, on the issue of interaction in combination product, the FDA states, "When combining products such as drugs or biologics and devices that are customarily developed using different regulatory paradigms, certain critical developmental issues, such as the interaction of the drug/biologic and device constituents, may not be readily apparent." On the issue of preclinical development, the FDA states:

> For example, guidance for preclinical evaluation of drugs/biologics differs from the preclinical/non-clinical studies conducted for devices. When developing a combination product, it is likely that neither isolated approach would fully address the relevant preclinical development questions for both constituents as well as for the combination product as a whole. Instead, FDA recommends that developers consider the scientific and technical issues raised by the combination product and its constituents and propose an approach that appropriately addresses these issues without requiring duplicative or redundant studies.

On the issue of technology used to develop combination products, the FDA states:

> Innovative new technology may also challenge existing approaches for product development. For example, a new device used to deliver a drug/biologic to a new area of the body that was previously inaccessible might make it necessary to develop new methods to determine the effect of such localized/targeted delivery, particularly when it results in higher exposure to that target than when the drug is systemically administered. Likewise, innovative technologies such as nanotechnology or live cellular products may lead to the development of new manufacturing methodologies or unique safety issues not associated with products manufactured in other ways.

All these clearly highlight the importance of developing a design and development plan.

As the Greek philosopher Plato said, "The beginning is the most important part of the work." It may not be the most fun part of developing a combination product, but it is certainly beneficial. There are many planning methods available today to the project leader, ranging from simple spreadsheets to complex Monte Carlo simulation techniques applied to resource planning. It would be ideal if activities that need to take place in all six domains in the road map can be planned in advance. But in reality, the design and development plan should, as a minimum, include activities in the middle four domains: customer, functional, design, and process. A typical plan focuses on including the following for each domain:

- Activities, start date, and duration for each of those activities and owners
- Resources and their loading (part-time, full-time, 20%, etc.) and other nonrecurring development costs needed to complete these activities
- Anticipated risks and associated mitigation plan

Regardless of the planning methods used, a project leader must include certain key activities in the plan. These activities include: design and development, risk management, regulatory, clinical, supply chain and manufacturing, quality, and postmarket service. For combination products, the plan should clearly focus on the definition of the device and the regulatory pathway, as this one aspect alone can potentially accelerate/delay product launch. We will be discussing regulatory aspects of combination products in Chapter 4.

In addition to basic project planning, Monte Carlo simulation techniques are useful in answering questions such as: How do I optimize the project schedule? How do I optimize the project budget? How do I optimize project resources? Software such as Crystal Ball can help the reader with simulation-based project optimization work. For combination products, it is highly recommended that the project leader goes beyond basic project plans to try and use advanced techniques for project planning as well as a more comprehensive approach to design and development planning. It can help the team and the company to minimize any schedule attainment risks.

Product development in healthcare product industries

Innovation domain

Activities in this domain, for a combination product development, will typically focus on identifying new opportunities for existing drug/biologic molecular entities using medical device/diagnostic instruments as a platform rather than activities that typically take place to discover new drug/biologic molecular entities. A quick look at the combination products that have been recently approved by the FDA (presented in Table 2.1 in Chapter 2) reveals this trend at least for the time being. The primary reason, we believe, are the complexities that companies in this space must tackle to develop a combination product with new drug/biologic entity and lack of capabilities that exist in people, processes, and other resources.

What does developing a new drug/biologic entity entail? Without going into the details, here is a summary of the discovery phase in pharmaceutical/biologic companies. Discovery is a firmly embedded practice in pharmaceutical/biologic companies where biologists and medicinal/protein chemists create new molecular entities (NMEs) and then recommend these entities for further development if they can satisfy unmet needs. One can easily imagine the efforts put forth in laboratories (internal or external to the company) by scientists with significant expertise and experience in a particular therapeutic area of interest. For example, pharmaceutical companies must develop an assay to test the efficacy of their drugs if they ever want to have a chance to get their drug characterized, and thus have their new drug application approved. These innovation domain activities result in deliverables such as Institutional Animal Care and Use Committee (IACUC) approval of all animal protocols, analytical study reports, a pharmacological profile, a compound monograph, and so on. It is well known that for every successful molecular entity that enters full development, there are many that fail. The costs and complexities are enormous.

If a company developing combination products does indeed develop a new entity to be a part of a combination product, it needs to take into consideration the potential interaction between the target pharmaceutical/biologic compound and the device/diagnostic element. This interaction needs to be adequately addressed during the innovation domain to avoid any surprises later in development. In addition, if the pharmaceutical/biologic agent's exposure to the device and the dose range are different than what is established, more work needs to be done in this domain to fully understand the interactions.

Device companies developing combination products (combination DP or DB in Table 3.1) must perform the following activities in the innovation domain:

- Complete technology and capabilities assessment to ensure understanding of pharmaceutical/biologic agent
- Evaluate the impact of the technology/processing on the active drug/biologic ingredient and understand how to apply the drug to the device
- Understand the analytical and assay test methods and consider the stability testing and strategy
- Understand the toxicology, pharmacology, and safety performance of the molecule
- Assess possible interaction between the drug/biologic entity and device parts
- Assess facilities, utilities, and environmental needs of the potential drug
- Evaluate the skill sets to address the regulatory, analytical, quality, and clinical needs of the combination product project

In the case of drug-coated stents, the typical questions at this stage would be: Can the proposed drug-device combination product have a better safety profile than medical therapy? Which pharmaceutical ingredient is appropriate to be coated on to the stent?

Drug companies developing combination products with a biologic (combination PB in Table 3.1) must perform the following activities in the innovation domain:

- Define and validate a research target through development of assays and surrogate animal models (as appropriate), and complete *in vitro* and *in vivo* studies leading to the identification and selection of candidate NME.
- Develop assays and animal models to human target, and understand the biophysical properties and potential disease state indications of the candidate NME.
- Conduct preclinical studies to support potential disease state indications and early pharmacokinetic studies.
- Finalize the candidate NME, secure intellectual property, and gain approval of appropriate governing bodies to move to early-stage development.

If the discovery part of the R&D work, or the development of technology and science, has been completed and the technologies involved in the potential combination product concepts have shown to be feasible (e.g., robust, measurable, controllable, predictable, understandable, and the clinical effect reproducible), then the focus during the innovation domain is mainly on the specific combination product application. For example, in the case of drug-eluting stents, the emphasis in the innovation domain would be on selecting the correct drug (cytotoxic vs. cytostatic, etc.) among the many options available, selecting the correct dosage profile, looking at various manufacturing technologies to coat the drug on the stent, focusing on securing intellectual property rights, bringing in the right talent for future development, and so forth. If that is not the case, then the bulk of the efforts in this domain will be on ensuring that the technologies involved in the potential combination product concepts are feasible (e.g., robust, measurable, controllable, predictable, understandable, and the clinical effect reproducible). For example, in the case of drug-eluting stents, the focus will be on the feasibility of technologies that can be used to coat the drug on the stent, which can ensure desired drug elution rates, and so forth.

Other activities in the innovation domain for combination products include:

- Production of drug substance in sufficient quantities with the desired quality using methods that include reagents, solvents, and conditions not suitable for scale-up and commercial synthesis.
- Performing basic compatibility studies to characterize the active pharmaceutical ingredient (if applicable) and identify conditions, excipients, and materials to select. In addition, it includes the study of interactions between the device and the drug.

- Early development of test methods focused on analytical chemistry as well as physical and mechanical properties of the combination product. It must be noted that the competencies and infrastructure are different for analytical chemistry and physical/mechanical test methods.
- Exploratory drug stability to determine the extent to which a drug substance retains the same properties and characteristics that it possessed at the time of its manufacture throughout its period of storage and use. If a biologic entity is part of the combination product, special attention must be given, as biologics are very labile substances and their stability and effectiveness can easily be altered by temperature, agitation, contact materials, and so on.
- Early focus on necessary modifications to the biological product, including modifications to the gene encoding process, expression/secretion of the product, and so on, to make it compatible with the medical device.

Contrary to popular belief, DFSS tools such as design of experiments can certainly help the scientists and engineers plan and execute their innovation efforts better. Once the company gets a better understanding of the science and technology of the combination product that is to be developed further, it can start focusing on defining the potential opportunity for the product. In the previous sentence, we deliberately indicated "once the company gets a better understanding" because we believe that for combination product development to be successful, it is not enough if only the scientists and engineers understand the product technology in the innovation domain.

Opportunity identification (innovation and customer domain)

Most medical devices and diagnostic equipment can be classified as tools, aid mechanisms, monitoring systems, or automated systems for diagnosis and treatment. Most pharmaceutical/biologic products, on the other hand, can be classified as therapeutic or chemical agents that act on specific organs. The challenge to the combination product developer is to identify how the tools or monitoring systems will be used to deliver therapeutic or chemical agents to act on a specific organ. This activity is called identifying customer needs, and it requires the mindset of the tool user in the specific trade. It is analogous to understanding a plumber's mindset when she or he is facing a clogged pipe that needs to be cleaned. The plumber can use chemicals, a wire brush, or a combination, depending on the location and the extent of the clogged pipe. If the plumber decides to use a combination, then he or she must know the type of chemical and the brush to use, in what sequence to use them, an so on. Though this is an overly simplified analogy, it clearly illustrates the need to understand the specific medical practice so that the process used to collect customer wants and needs (e.g., surveys) can intelligently assist during the innovation domain. In some cases, it would require combination product developer(s) to anticipate or speculate the future of medical practice. For example, in the case of drug-eluting stents, it would require that the product developer speculate how (and how long) the drug and the stent will be delivered to the clogged artery without any undesirable side effects. In cases such as this, statistically justified customer surveys or interviews may not provide the required input for design. The voice of the customer (VOC) at this stage can be purely speculative or "over the horizon," but it is partially useful to define what is needed. How, then, would one go about collecting the customer voice? One way to do so is to make sure that the potential customers are fielded open-ended questions. The reason for this is that statistical methods were invented to draw conclusions from static populations, while innovations such as combination products imply the creation of such a population. The combination product area is such a dynamic one, and therefore is full of uncertainties—technical and business. The technical uncertainties are comprised of clinical effectiveness, acceptance by the medical community, clinical cost effectiveness, and manufacturing cost effectiveness. The business uncertainties for combination products include unknown regulatory pathway and revenue targets. Instead of using customer data to interpolate (e.g., draw a conclusion based on the sample), the data will be used to extrapolate (e.g., establish a hypothesis and set up an experimental plan). In many cases, all that is needed is the attention of a leading surgeon/healthcare provider who, once the device is developed, can influence others in his or her area of specialty. Regardless of the approach taken, activities in the innovation domain are focused on identifying and defining the business opportunity based on a value proposition to the healthcare community and patients. Therefore, these activities are not to be regulated by the FDA or any other regulatory agency.

Customer domain

The traditional product development process in pharmaceutical and biologic industries is more of a technology push approach than a market/customer pull approach. What do we mean by that? Technology push purely focuses on identifying, developing, and launching a drug/biologic product after clinical trials to the marketplace. The customer groups for the drug/biologic product (e.g., internal medicine specialists) would rarely be interviewed up front during the development process. A market/customer pull approach, typically used in the development of medical devices, on the other hand, tries to involve customer groups up front in the development process. Since most of the combination products approved so far are based on medical device platforms, it would be safe to suggest that a hybrid approach that includes market/customer pull would be beneficial to the development of combination products.

In the customer domain for combination products, the first activity a project team should focus on is to assess the variability (beliefs and perceptions) among the healthcare professionals in using the product being developed.

For example, gastroenterologists that perform endoscopic procedures deal with certain patients that develop a narrowing of their esophagus called strictures. These conditions require a mechanical procedure called esophageal dilation (reopening of the tube). One school of thought believes in ignoring the duration of stretching the esophageal tissue, while the other school of thought believes in a 60-second minimum stretch. Another example is the variability in the number and size of drug-eluting stents to be used to prop a clogged artery open. While some interventional cardiologists prefer deploying two smaller stents, others may prefer one larger stent for the same occlusion length. Our intent here is not to identify which practice is correct. Instead, our position is to emphasize that knowing the existence of both is relevant to the product development team. This approach will help the team in defining the combination product's reliability, and so forth, appropriately. In the first example mentioned, the device's mission profile can be defined assuming the worst case (e.g., 2 × 60 = 120 seconds esophageal tissue stretch, with a safety factor of 2). Even in the second example, the device's mission can be defined based on the worst case (e.g., two smaller stents).

We introduced the term *VOC* (voice of the customer) earlier. This term is used largely by Six Sigma professionals, but it is applicable in combination product development as well. VOC is an approach to effectively deal with the variability mentioned in the previous paragraph. Combination product reliability planning is totally dependent on VOC, and it starts this early in the new product development (NPD) process. The VOC approach also lays the foundation to define the market segmentation or development of a marketing strategy. It also helps avoid overreactions to contradictory customer input. How often have you heard a marketing person dictate terms to the product development team as to customer wants and needs? It is this overreaction that causes product development teams to lose focus and become diluted in trying to meet the must-have needs, nice-to-have needs, and delighter needs from all potential market segments. As a result, the project development schedule can become as stretched as the esophageal tissue during surgery.

The fallacy that only marketing or business development groups are supposed to know and be accountable for the customer domain inputs is not optimum and is certainly not in line with DFSS principles. It the responsibility of the project team leader (for that matter, the president or CEO of the company) to ensure that the entire combination product development team fully understands customer wants and needs as well as appropriate clinical procedural knowledge prior to the start of any prototype development work. This activity cannot be emphasized enough, as it is a cornerstone for any effective product development effort.

The project charter

Earlier in this chapter, we mentioned the need for establishing a development plan. A project charter is a document that we recommend project teams develop after the innovation domain activities are complete. This document defines the objectives of the combination product development project and includes the scope of the project. Typically a function such as business development or strategic marketing is the driver and main participant of the data gathering to creating the project charter. Questions such as those listed below are typically answered prior to the project charter creation:

- What market segments are we targeting with the combination product? (e.g., outpatient clinics vs. hospital)
- Will this combination product meet an existing procedure need or create a totally new medical procedure? (e.g., based on published literature, it can be seen that the launch of the drug-coated stent did not require specialized training in deploying the stent but required training on the use of blood clot prevention techniques)
- Will this combination product be launched globally or in the United States only? (e.g., compatibility issues with Japan or European hospitals vs. U.S. hospitals)
- What will be the primary mode of action (PMOA) for this combination product? (e.g., Which FDA center will have primary control over the regulatory approval application for this product—CDRH, CBER, or CDER?)
- What will be the appropriate clinical research strategy? If clinical trials are held outside the United states, which locations would be ideal to generate relevant data?
- What is the product pricing strategy? Will the insurers offer reimbursement for the product?
- What are the risks of this project?
    - Technological (e.g., first time the technology is used in the marketplace)
    - Regulatory (e.g., a first-to-market product with no predicate product; may need to generate clinical data)
    - Business (e.g., What if a competitor or a small, relatively unknown company comes up with a better concept to the market first?)
    - Manufacturing (e.g., What if the manufacturing factory cannot make a large volume of this product due to new technology?)
    - Financial (e.g., What if the request for large initial capital investment may sound too risky?)
    - Schedule (e.g., What if management underestimates the need for additional time for project completion due to product complexity?)
    - Resources (e.g., What if there are no adequate and competent resources available to work on this project?)

While a cross-functional product development team may not be needed to provide adequate answers to these questions, they can certainly help in situations where either the combination product technology is new to the company or the team members are new to the project, or both. It must be mentioned that while a project charter is typically a formal business document, a design and development plan is a formal regulatory document that can be audited by regulatory authorities or notified bodies.

Customer wants and needs: *Design inputs (customer domain)*

At this stage in the development process, the project team should have all the necessary technical knowledge and background for the combination product. They may not, however, have comprehensive data on customer wants and needs. For example, in the case of drug-eluting stents, the team might have data on the safety and toxicology profile of the drug (Taxus/Sirolimus) but may not have data on the preferred stent sizes, stent profile, and so on. The VOC collection process is an iterative one, just like product design and development or problem-solving processes. The team goes through this process until a set of refined statements can be made about what needs to be designed (e.g., intended use).

The design and development team should have continuous but well-planned access to the customer. Design input for the combination product includes inputs on the physical characteristics of the product, patient educational materials, product shelf life and handling requirements, product life, intended functions, labeling, servicing, components, accessories, and so on.

As we stated earlier, many open-ended questions will be raised with different customer segments in the beginning. In addition, after the proof-of-concept testing is successful, the design and development plan typically will need more VOC to be collected to refine data collected earlier. For example, it might be necessary to find out how the proposed combination product will be distributed in a hospital setting or what educational materials might be needed to support customer (or their family members') education. Once the design and development team develops early prototypes to show to the customer, additional customer visits might be required. At this stage the questions will be more targeted at finding specific answers, and to set a more accurate and specific direction. In our experience, most healthcare professionals need to see and feel the proposed product in order to provide any useful feedback. By going to other potential customers at this stage than the ones utilized earlier, the product development teams might increase their sample size, but this may not be the best strategy since there is no baseline reference to indicate that the teams are moving in the right direction based on customer input (experimental confounding effect). For example, if the product development team collects VOC in the beginning of the project from a 60-year-old interventional cardiologist that resists adopting the stent if she has been using balloon angioplasty successfully for the last 25 years, then it would be wise if the team goes back to her when they have a working prototype instead of showing it to another interventional cardiologist with whom they talked before. Worse yet, if the second interventional cardiologist is used to bare metal stent angioplasty, then he or she may confound the VOC analysis. It is also possible that the 60-year-old interventional cardiologist may never be willing to switch to drug-eluting stents.

Since combination products are complex in nature, it is quite possible that the product development teams might have to develop one or more prototypes to gain full understanding of customer wants and needs. This is because most surgeons and other physicians can state their likes and dislikes much better if they can feel, see, and touch the combination product prototype. One can ask if showing a three-dimensional computer-generated model is good enough, but we have seen in many cases that it is not. Thus, one can clearly see that VOC is an iterative process that starts with gathering initial customer voice, developing prototypes, collecting additional VOC to fine-tune the prototype, and so on.

We strongly encourage the combination product development team to adapt/adopt the following tips for activities planned in the customer domain:

- Ensure that every team member has access to the customer at all times during the product design cycle (not just the marketing representative) in order to develop a comprehensive list of requirements. For the drug-eluting stent, this could mean that stent design engineers, drug scientists, and so on, have access to the customer to develop requirements for drug elution rates, and so forth.
- Ensure that all team members are knowledgeable in areas associated with the combination product to be developed. These areas include knowledge of the human body (e.g., anatomy, hematology, immunology, etc.), the medical treatment, the therapeutic path, and the scientific principles involved in the predicate device and its interaction with the patient and the user. In the case of coronary drug-eluting stent design, this could mean that the design team is knowledgeable in angioplasty procedures, coronary arteries and their function, coronary artery disease (CAD) and its effects, anti-blood-clotting therapy, and so on.
- Allow the customers to lead the conversations instead of leading the customers, while paying attention to their environment and unspoken words. In the case of coronary drug-eluting stent design, this means allowing interventional cardiologists to lead the conversations and the design team paying attention to things like frequency of anticlotting drug use, sizes and types of products used to treat CAD, and types and conditions of patients treated.
- Document VOC sessions to effectively guide the design and development efforts and to show the FDA and other regulatory bodies that design input efforts were meaningful. Defaulting to the documented VOC is the easiest way to solve the typical design project disputes that normally arise among team members. Many a times we have heard marketing experts and design engineers argue why a particular feature

or function is important (or not). In those instances we have seen VOC documents help settle the arguments.

- Understand the customers' preference to products with health insurance reimbursement policies. For example, in the case of coronary drug-eluting stents, it is important that the product development team knows whether the majority of interventional cardiologists would prefer a drug-eluting stent if they knew the product has reimbursement from the insurance providers. In addition, it is also important for the team members to know that lifesaving therapies are generally considered good candidates for reimbursement in the United States, but the team has to show clear clinical evidence since interventional cardiologists will tend to use drug-eluting stents if they know that there is clear and positive clinical evidence that speaks to the benefits and risks of their use.
- Surveys, interviews, focus groups, and ethnography are the tools of choice for market research in the world of medical devices. These techniques can be utilized in the case of combination product development as well.

Table 3.2 depicts some examples of typical wants and needs that can be expected from customers in the world of combination products. Note that similar to the auto, personal computers, and consumer electronics industries, it is not only the combination product that matters to the customer, but also the associated services and human contacts.

Table 3.2 Typical Voices of the Customer in the Healthcare Product Industry

VOC	Patient	Doctor	Nurse	Purchasing agent
VOC1	Needs to be safe, should be covered by insurance, result in minimum pain with no side effects, etc.	Needs to be easy to use/intuitive, have less morbidity/mortality and excellent and proven clinical benefits	Easy to use, store, unwrap, dispose, clean and sterilize; minimal nurse supervision of the patient	Cost-effective, reimbursable, excellent product-related service, reputed company
VOC2	Should have no pain, no allergic reactions, and be able to walk for miles	Easy to place during procedure, excellent support from sales rep, overwhelming clinical superiority	Easy to use, clear instructions with plenty of pictures, ergonomic and intuitive	Cost-effective, reimbursable, excellent product-related service, clear codes that can easily be entered in system

Once customer wants and needs are collected, it is important that the combination product development teams identify product design requirements. This includes user needs and intended uses from the drug or biologic component. Typically these will include development targets such as elution, pH, and so on. In addition to user needs, combination product teams must gather requirements from internal customers such as manufacturing operations and packaging development. These requirements will include targets such as process capabilities, manufacturing location, and so forth. Oftentimes, the teams must assess the internal customer-based design requirements against external customer-based design requirements. Design for Six Sigma (DFSS) tools such as quality function deployment (QFD) can help the team effectively complete this assessment and create a prioritized set of design requirements. We cannot emphasize the importance of developing these design requirements prior to any further action. Teams can sometimes tend to assign design requirement targets as TBD (to be determined) at this stage. This is an acceptable practice only if requirements cannot be established until prototypes are developed. Otherwise, it is not acceptable and can even land the teams in a disaster later, especially during verification or validation stages.

After establishing design requirements, the next immediate step for the teams is to perform an assessment of the capabilities available to meet these requirements. These capabilities include technology, manufacturing, analytical, and test methods, stability testing, people competencies, facilities, and so on. Questions such as "How should the drug be applied to the device?" and "What should we do to understand the toxicology, pharmacology, and safety performance of the molecule?" need to be answered.

Concept selection

Once design requirements are established and capability assessments are performed, it is time to develop concepts. Most of the practitioners in the industry know how hard it is to develop meaningful concepts. Many items need to be addressed to result in one or more concepts that can provide the team with enough assurance that the team is moving in the right direction. These items include, but are not limited to, intellectual property review, pharmacological profile, safety profile, competitive profile, device/drug/biologic compatibility, and product stability. For a drug-coated stent concept, questions such as "Is the stent material compatible with the polymer selected for drug coating?" and "How do we ensure that the drug is stable?" can arise. In general, for combination drug-device products, studies that are carried out in the pharmaceutical industries, such as preformulation studies to support phase 1/phase 2 formulations, provide clear direction in selecting concepts that can help kick-start the commercial product design. The scalable drug formulation and sterilization methods to be used are typically known at this point. The process of concept generation, evaluation, and selection is iterative in nature. Once the team feels that they have a combination product configuration that can potentially meet VOC and design requirements, they are usually ready to move to the next step. However, concepts must be selected using DFSS tools such as the prioritization matrix or Pugh matrix, as these tools can help the team evaluate concepts based on not only external customer requirements but also internal requirements.

Before the project team dives deeper into customer input-related topics in the customer domain, it must complete proof-of-concept tests.

Proof-of-concept testing

Toward the end of the customer domain or the beginning of the functional domain, after ensuring all the key innovation domain activities mentioned above are completed, the combination product research teams perform one or more proof-of-concept tests to clearly understand the human physiology and mechanism of action for the product, and to ensure a high degree of confidence in the predictability of the animal models. Proof-of-concept testing is important for the following reasons:

- It signifies that the combination product idea is backed by proper business valuations, has sufficient intellectual property protection, and has a business plan in place for clinical trials, and so forth, if necessary.
- Successful completion of proof-of-concept testing signifies a clear hand-off from the research team to a clinical or commercialization team. In addition, it also indicates that there is a possibility that the idea can become a successful medical treatment.

Figure 3.5 is a simple five-step proof-of-concept process that can be applied to combination product development. For example, in the case of coronary drug-eluting stents, the first step is to identify the environment—blocked artery with limited blood flow. The next step is to set up a test environment capable of emulating the human coronary artery. Typically these are animal models that are used for testing the proposed drug-eluting stent concept. These animal models are monitored at regular time intervals for effective prevention of restenosis. Step 4 is to test various types of drugs and stent size

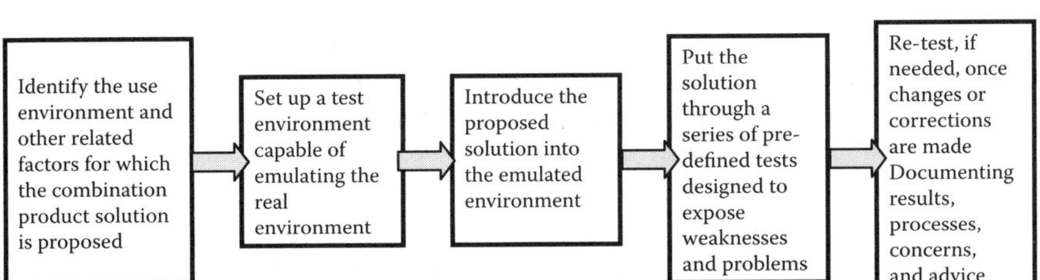

Figure 3.5 Proof-of-concept testing process.

combinations. Six Sigma tools such as design of experiments can be effectively applied to minimize the number of experimental runs without compromising the quality of test results. Problems, if any, such as stent breakage and difficulty in stent placement, will be observed at this step. Finally, the tests are repeated with changes and the results are documented.

Proof-of-concept testing is sometimes performed at the end of the innovation phase. This may be acceptable in a situation where the technology has little need for customer input, but we rarely come across such situations in combination product development. It is our opinion that it is performed at the end of the innovation domain primarily to satisfy organizational need to "touch and feel" a concept, and also to satisfy management desire to see progress. Concepts are better developed if the team understands customer wants and needs for a combination product.

Risk analysis and management

Formal risk analysis and management is a fairly new practice in pharmaceutical and biologics industries. Fairly new standards and guidances such as ICH Q9, *Quality Risk Management*, and the FDA's *Guidance for Industry Premarketing Risk Assessment* and *Guidance for Industry Development and Use of Risk Minimization Action Plans* help these industries in effectively addressing product risks. Risk management is intended to identify hazards and harms associated with the product, estimate and evaluate the associated risks, implement risk control measures, and monitor the effectiveness of the measures. Note that risk management should be managed throughout the product development and commercialization process. Risk analysis should be started as early as possible, even in the innovation domain. In fact, for some devices, effective risk mitigation is the actual value proposition. For example, according to a corporate press release (February 23, 2004, at www.BD.com) from Becton, Dickinson and Company,

> the BD.id System is the first to fully integrate bar-coding technology with proven specimen collection process standards. The system enhances patient safety by helping to reduce the potential for errors during the specimen management process. Specifically, the system helps to ensure that blood and other samples are collected from the right patient, are placed in the proper container and are labeled correctly.

For a drug-device combination product, we recommend that both ISO 14971 and ICH Q9 are used to conduct risk management activities. It is necessary that we mention that new risks (e.g., patient death, adverse reactions, etc.) associated with combination products that may not have been associated with either the drug or the device separately might be identified. Regardless of the association of the risks (product, patient, etc.), it is important that all risks be appropriately considered via risk management throughout the life of the product.

Risk analysis and management should be a multidisciplinary approach, but it depends on the type of risk being addressed (clinical application, product design, or manufacturing process). Our recommendation for the teams is to use a facilitator as much as possible during risk analysis sessions early in the development process. The team members responsible for this activity should have knowledge of standards (ICH Q9, etc.) and any appropriate company policies/procedures. For a drug-biologic product, it must be mentioned that biologics and materials of animal origin have strict risk standards for use.

We also recommend that companies set up appropriate standards to make decisions on whether to reduce or accept risks. If DFSS tools such as failure mode and effects analysis (FMEA) are used to perform risk analysis, we recommend that the standards for risk occurrence include both qualitative and quantitative scales. For additional information on the FMEA tool, refer to other books by one of the coauthors published by CRC Press.

Functional domain

Activities in this domain, for combination product development, will typically focus on defining the various functions and functional requirements that the product should meet using prototypes. Team members with device and drug/biologic expertise typically will work together to determine what the combination product is supposed to do. They will also determine what the users of the combination product are supposed to do make the product work. Together, they are the two key aspects of functional design: system functionality and user functionality. For a combination product to be successful in both clinical trials and commercial use, the development team must effectively bring these two aspects together. In order to do that, it is important that the designers or design engineers develop a formal document, usually called functional specification. This document is nothing but a collection and analysis of all of the functional requirements for the combination products in both aspects mentioned above. The analysis of the requirements starts with the overall system description. This is then drilled down (broken down) into subsystem elements, thereby allowing the creation of hierarchical and manageable functional elements. The level of detail also increases from the top system level to the bottom component level. Care must be exercised when creating functional specification for combination products. Since these specifications are so focused on the combination product technology, the teams must ensure that (1) the VOC collected earlier is clearly linked to design requirements using DFSS tools such as QFD, and (2) these design requirements are clearly linked to the functional specifications. If not, these functional specifications can lose their context and might even result in a product that is fully functional but does not meet customer wants and needs. Since the development teams often rely on these functional requirements, it is very easy to see that these can promote a checklist mindset, leaving the team feeling too confident that they have achieved the needed results.

It is important in functional design of a combination product to determine which decisions are made by the product and which decisions the user should make. For example, in the case of drug-coated stent, stent diameter, the amount of drug coated, and the amount of sterilization are all predetermined by the stent, thereby providing consistency without any input from the user. However, choices and options for the user, such as stent size, are still integrated into the design but are made clear to the user.

At this stage in the design of the combination product, the teams should have clear knowledge of the following:

- A functional prototype that needs to be designed further
- A clear high-level understanding of the risk posed by intended and unintended uses of the product
- A clear understanding of the requirements that are linked to customer wants and needs

The next step is to effectively manage design requirements so that they flow down along with the design all the way down to the component level and then on to the process level.

Management of design requirements (from initial design inputs to final design outputs)

Management of requirements is a familiar concept in defense and high-technology companies, since they have been utilizing it since the early 1980s, especially with the advent of relational databases. The clear need to define what is needed before delivering results, similar to the FDA's design control guidance, where inputs come before outputs, is fundamental to this concept. This principle is embedded in the quality system regulations (QSR) design control requirements, ISO 9001 and ISO 13485. However, we are all too familiar with product development teams' (due to pressure from management) tendencies to go for a shotgun approach, where the teams tend to design first and then establish requirements using a reverse engineering approach after they have a functional prototype. This approach works sometimes, especially for small start-ups with a great idea and a handful of team members. When one comes across a team that has a design prototype that somewhat works but a lot of TBD in their design requirements, one can clearly jump to the conclusion that DFSS principles were not followed because the designer has skipped over the define, measure, and analyze steps in the DFSS road map—define, measure, analyze, design, verify and validate (DMADV)—and may have just superficially touched the innovation, customer, and functional domains. This is troublesome especially in cross-functional product development team structures, where nondesigners will have to sit and wait until the after-the-fact requirement documents are generated (e.g., design inputs and specifications).

To the contrary, a formal set of customer and design requirements can become a common target to all team members regardless of the specific skills that they bring to the project. The DFSS way of requirements, cascade-down, is depicted in Table 3.3. This table will make sense to a systems engineer who has been educated in areas such as Request for Proposals (RFP), proposal creation, proposal evaluation, systems integration and testing, systems commissioning, and so on. In general, we can say that many electrical engineers

Table 3.3 DFSS Requirements Cascade

Design inputs, flow-down	(Potential) customer wants and needs	Practical interpretation	Design outputs, flow-up
	High-level system requirements	How should the product look in order to satisfy customer wants and needs?	
What other nonunique or old requirements are needed?	Systems assembly, final configuration, and final acceptance specifications		
	Subsystem or subassembly functional requirements (technical design inputs)	What engineering or technical schemes are needed to make the product?	Subassembly, subsystems assembly, and testing specifications
	Component design requirements	What components are needed? What is the role of the component regarding function of the device? What physical, chemical, or biological characteristics are needed for this component? What is the allocated reliability to this component? Why?	Component acceptance specifications
	Production requirements	How can we manufacture? What technology is needed to satisfy quality, cost, and service level?	Component manufacturing and process control specifications

and certainly software and computer engineers are formally trained to work this way. However, it is important for other team members and management to be aware of differences among the technical backgrounds. It must be mentioned that the field of requirements management is related but not equal to critical parameter management.

A systems engineering approach to management of requirements

In the mid 1980s, high-technology companies like AT&T Bell Laboratories came with company-wide initiatives such as systems engineering and the creation and adoption of a program called Design for Excellence (DfX; pronounced "design for x"). The systems engineering role at Bell Labs was fascinating in all senses of the word: a truly diverse group of engineers with different educational backgrounds working together with the goal of defining the high-level system requirements that could define a telecommunications system capable of meeting future needs and trends of the customer. The key thing here is that this group of systems engineers did not know much about electricity, radio frequency, switching, transmission, and so on, but were very good at understanding customers and in using logic to define their needs or future wants (VOC or design inputs). A second group of systems engineers would then take those requirements and high-level system architectures and would try to define how to meet the needs with current technology and what was the new technology needed to complete the set of customer needs (low-level system requirements or more specific technical inputs). The key element back then, and, for that matter, today's DFSS methodology, is thus management of requirements. In other words, how are the requirements flown down or cascaded during the design and development process? The availability of relational databases and other information technology tools will also help to manage the configuration of complex combination product systems during design and development. Design changes occur throughout the entire design and development cycle, and therefore, it is important that, at a given point in time, the team members know if the elements of design that had already been approved are affected by a change in the components. Similarly, they should also know answers to questions such as "Where did the tolerance or the specification come from?" and "Who among the team members can answer specific questions if there are 50 scientists/engineers developing a combination product across three continents?"

The other initiative at Bell Labs was DfX. This initiative implied the need to consider additional goals aimed at the internal company customers to any given state-of-the-art design initiative. For example:

- Design for manufacturability
- Design for reliability
- Design for maintainability
- Design for testability
- Design for simplicity

These design goals became an integral part of a successful product design. It was no more about making the design work, but how to systematically get to successful design with dozens of highly technical and business people who are very specialized in their areas of expertise. These ideas were also combined with Taguchi's concepts of robust design, the loss function, and parameter design (transfer function).

In the medical device, pharmaceutical, and biologics industries, but especially in the medical device industry, the evolution toward a systems-based approach started when the traditionally bureaucratic, audit-oriented, functional-quality professionals became an integral part of the cross-functional product development teams. In their new role, these quality professionals were asked to understand and positively contribute to the design and development of the product and the associated manufacturing process. As a result, these professionals had a narrower but more important and relevant role as well as accountability for project results. The next evolution was adding regulatory knowledge to the list of their competencies so that they could answer questions posed during FDA or other approval agencies' inspection. With the advent of DFSS methodology in the last decade, these professionals were asked to evolve into black belts or master black belts in product development by adding more systematic problem-solving methodologies and business knowledge to their skill set. And of late, as product development in these industries is becoming much more complex and companies are forced to be more "lean" to improve productivity, these professionals are being asked to become systems integrators or systems engineers. While it is extremely difficult to find individuals with all these competencies, it is imperative to evolve to such cross-functional, multidimensional roles to enable successful development of combination products.

This evolution is not limited to quality professionals but includes all functions that need to support combination product development. Whatever the name of the position is, the need of the business is simply having well-understood customer, functional, design, and process requirements that are integrated into the design and also well documented. If these needs of the business are met and the right disciplines are in place, the combination product development team and, for that matter, the industry should have no issues with meeting the regulatory requirements. While smaller companies developing combination products may have cross-functional, multidimensional roles due to necessity, it is important for senior management in large companies to ensure its presence in combination product development.

Initiating the requirements cascade (from the customer domain to the functional domain)

Once the customer requirements are collected it is important that right VOC from the customers or potential customers is correctly translated to system requirements. The adequacy of the translation process is in fact very explicit in the medical device quality system regulation. It is imperative to realize that design inputs are a chain of requirements that may require unfolding or cascading many times. This means that the outputs from the first iteration become inputs to the next, and so on. The translation of customer requirements into functional requirements may take several iterations, and keeping track of such activities may be of benefit to the combination product development team. Note that the high-level system requirements here do not give any idea about how the final design will look. However, explicit measurable output, from a set of design parameters, has been defined as criteria to satisfy the VOC.

Requirements cascade

According to the *Encarta World English Dictionary*, a cascade is a succession of things, for example, chemical reactions or elements in an electrical circuit, each of which *activates, affects, or determines* the next. In terms of requirements management, the cascading starts with user needs at the top of the pyramid and ends with the design details at the bottom. The shape of the pyramid suggests that there will be many detailed design characteristics or parameters, for each customer need. The base of the pyramid also indicates the most fundamental principle of safe and effective design of medical devices engineering and scientific expertise. In our experience, developing a dependable and self-directed engineer or scientist in the healthcare product industry takes many years, maybe even a decade.

What is the difference between a requirement and a parameter?

A requirement is something that is needed for a particular purpose according to the Encarta encyclopedia. According to the *Webster's Dictionary*, a requirement is something essential to the existence or occurrence of something else. This latest definition will be very helpful to realize that the cascading of requirements into lower levels is what the "something else" means. For example, an immunoassay requires a sensitivity of 5%, and the only way to obtain such sensitivity is by designing an assay with a very steep calibration curve at the lowest concentrations of the assay. Thus, the calibration curve (a design parameter or variable[*]) becomes a design parameter from which the functional requirement *sensitivity* depends. Exposing the reagents and the instrument to a blank sample typically assesses sensitivity. If there is no analyte, then there is supposed to be zero reaction. But there is always something detected by the instrumentation that electrochemists would call noise or background. Thus, the average level of detection plus two standard deviations[†] is inverse regressed into the calibration curve as a concentration and then converted to a percentage. Interestingly, it is our experience that the mathematics and statistical elements of immunoassay development and

[*] Different levels of "steepness" will define different levels of sensitivity.
[†] As defined by the National Committee for Clinical Laboratory Standards (NCCLS).

control in the diagnostics industry are typically not well understood by the biochemist that designs the assay.

Companies that vigorously and diligently practice DFSS principles or DFSS groups within large companies recognize these competency gaps and ensure that scientific groups (e.g., personnel with life sciences background) are complemented with engineers. In fact, in Mikel Harry's original Six Sigma breakthrough strategy, instead of the five fundamental steps of DMAIC, there were eight steps, as in RDMAICSI. The *R* meant recognize. The definition of parameter is then any set of physical properties whose values determine the characteristics or behavior of something (*Webster's Dictionary*). A limiting factor is a fact or circumstance that restricts how something is done or can be done (*Encarta Dictionary*).

The most efficient way to explain the rest of the cascading and DFSS methodology is by illustrating an example of a simple device such an IV set.

IV Set Example

> An intravenous fluid set is a typical drug delivery system that consists, in its basic configuration, of a spike, drip chamber, clamp and roller clamp, and clear tubing. Table 3.4a illustrates the flow-down, while Table 3.4b illustrates the flow-up for this product.
>
> In order to create the flow-down properly, transfer functions must be established that help transfer requirements from high-level systems to lower-level subassembly/component design requirements. For example, if XS = extra-strong IV set, then the transfer function could be
>
> $$XS = f(BS)$$
>
> and
>
> $$BS = f(RMLL, SY)$$
>
> where BS = bond strength, RMLL = raw material for the luer lock, and SY = type of solvent Y.

What is functional?

Functional is when something is specially fitted or used, or for which a thing exists, or when a group of related functions contribute to a larger action (*Webster's Dictionary*). So, in the functional domain, requirements that describe the (potential) customer wants or needs are defined. In Juran's world, quality is said to be fitness for use. The functional requirements, also known as high-level systems requirements, describe what the combination product is supposed to do for the customers.

Table 3.4a Flow-Down Cascading for Extra-Strong IV Set

Domain mapping	Requirements cascading	Flow-down
1. (Potential) customer wants and needs	A suitable IV set is needed for ambulances and trauma rooms.	
2. High-level system requirements—functional requirement	An extra-strong IV set should have the same functionality, capabilities, and dimensions that a normal IV set has, but with a special strength and a special labeling, indicating it to be extra strong.	Extra strong is the functional requirement.
3. Subsystem or subassembly functional requirements (technical design inputs—design parameters)	The bond strength between a luer lock and the tubing should withstand p pounds of axial force without detaching from the tubing. A 99% reliability level at 95% confidence for a safety factor of 3 is required.	Bond strength is the main design parameter.[a]
4. Component design requirements. (The raw material and the solvent are a response to the bond strength; thus, they are a design output aimed at defining the bond strength.)	The raw material for the luer lock will be X and the solvent Y.	Raw material and the kind of solvent are lower-level design parameters.
5. Production requirements (design for manufacturability)	The luer lock will be made by plastic injection molding controlling the following process parameters: screw speed, barrel temperature, and mold temperature.	Screw speed, barrel temperature, and mold temperature are the process parameters that dictate the output.

[a] Parameter implies variable, something that, by changing its value, changes the outcome. In this case, the outcome is the design.

From the functional domain to the design and process domains

Once the functions are defined for a combination product that is being designed, it is important to start activities that will help characterize the design or product output. What does that mean? The team has to identify, quantify, and control specific design parameters of a combination product so that the final design output meets design inputs regardless of the output produced or processed. It is important that the product and process are adequately characterized prior to design verification and validation. Why should a team do that? Because developing a clear and fundamental understanding of the relationships between design inputs and outputs will result

Table 3.4b Flow-Up Cascading for Extra-Strong IV Set

Domain mapping	Performance requirements	Flow-up (process and design capabilities)
6. Component manufacturing and process control specifications (design verification, design transfer, and process validation)	A process capability ratio of 1.33 is required for the entire operating range of the process parameters in injection molding of the luer lock.	The injection molding process is statistically stable, with a Cpk of 1.47 and a Ppk of 1.23 across the entire universe of possible combinations of process parameters (worst-case conditions).
7. Component acceptance specifications (design verification, design transfer, and process validation)	There are three relevant or critical dimensions in the luer lock that must be met at ±0.001 inch. These dimensions are length = 0.87 inch, I.D. at 0.11 inch, and O. D. at 0.16 inch.	All three critical dimensions where met, including all those devices made at the worst-case conditions.
8. Subassembly, subsystems assembly, and testing specifications (design verification and validation, design transfer, and process validation)	Before inserting the tubing into the luer lock, the solvent will be applied and a curing of T minutes will be allowed. The required bond strength is p pounds when stretched.	At 99% reliability and 95% confidence, a safety factor of 3 was obtained during a stress-strength test. Curing time is a design parameter that needs control in manufacturing. During the reliability test, it was realized that the connection luer lock to tubing is stronger than the elongation of the tube, which is the failure mode when stressed is applied. No new hazards are added by the elongation of the tubing.
9. Systems assembly, final configuration, and final acceptance specifications (design verification, design transfer, and process validation)	The extra-strong IV set has the same functionality, capabilities, and dimensions that a normal IV set has. Additionally, it has a special strength and a special labeling, indicating it to be extra strong.	The potential elongation of the tubing at p pounds does not jeopardize the intended use under worst-case conditions at 99% reliability/95% confidence.
10. Validated design with a validated manufacturing and assembly process	The DMR and the DHF are complete.	

in successful design verification and validation efforts. In the case of drug-eluting stent, this means that the team has to review and apply (where appropriate) the following:

- The impacts of various sterilization approaches on the drug ingredient
- The impact of analytical test method development since drug ingredient test methods are more stringent
- The impact of drug ingredient attributes such as impurities, solubility, and so on, on the performance of the product
- Technology transfer/scale-up (DP):
    - The scale-up process is significantly different for pharmaceuticals than for devices.
    - Scale-up efforts for pharmaceuticals are pronounced and need to be evaluated early.

If the combination product is a device-biologic product, it is important to note that there is a lot less flexibility in designing a molecule than a device, and minor modifications can occur in the device throughout the development cycle. However, controls in making changes on biologics diminish rapidly once a formula is designed. In addition, no animal-derived material should be used in the manufacture of biologics, and cell banks can be used as a source for testing/reference standards.

Design characterization is an important step in the design and development of combination products. DFSS techniques such as design of experiments (DOE) must be utilized to clearly establish relationships between design inputs and design outputs. This will enable the manufacturing or process development group to maintain the integrity of the design during the life of the product. Once the product is designed, it is important to perform tests to establish the product will perform as intended. Many types of tests are performed (reliability life tests, stability tests, shipping/transit tests, etc.) in this domain, but we will focus on a few key ones. We suggest that the reader refer to the publications at the end of this chapter for additional information on applying these techniques.

Stability studies

The intent of stability (aging) studies is to provide clear evidence on how the performance or quality of the combination product changes over time. These changes happen due to various reasons that include, but are not limited to, environmental factors such as temperature, humidity, and light, as well as shipping and handling factors such as storage and shipping conditions. For a thorough understanding of the combination product, the team may choose to perform stability studies on raw materials, components, and so forth.

- Device stability is of 6 months and is awarded on the basis of only pre-clinical data. Drug substance/product stability must be determined prior to any human clinical study.
- Early pharmaceutical/biologic clinical drug supply can have a rolling expiration date based on data from ongoing stability studies. Similarly, rolling expiry can apply for medical devices.
- It is critical that the medical device company understand the breadth and depth of pharmaceutical/biologics stability vis-à-vis the typical medical device aging studies. Medical device companies should be careful not to underestimate the resources (e.g., people, facilities, etc.) required for stability studies.
- Investigation for out-of-specification, out-of-trend, resampling, and retesting requirements is rigid in pharmaceuticals/biologics. Analytical laboratories should have documented procedures for reporting, recording, and rounding of results, retesting, and resampling. Trending analysis of laboratory investigations is required, and open corrective actions are followed until closure. Additionally, evaluation of corrective actions is performed to ensure that corrective actions put into place have been effective.
- Pharmaceutical/biologic companies typically file for registration with real-time stability data (minimum of 12 months).
- Stability testing on biologic device combination products is dictated by the biological component (biological stability testing is typically performed at 2°–8°C).
- It is not possible to use accelerated stability studies to project the shelf life of a biologic at the mandated storage condition, because it is not possible to apply the Arrhenius equation for degradation of biologics. Real-time studies are required to establish shelf life. Although it is not possible to project shelf life from accelerated studies, such studies are nonetheless performed to characterize product changes over time.

In our experience we have observed that successful launch of many products, regardless of whether they are combination products, is directly correlated with the success of the stability studies.

Design domain

The purpose of activities in this domain is to identify, quantify, and control the specific design parameters of a design element so that the final design output (product and process) meets technical design inputs. Most of the work done in this domain will be an extension of the work done in the functional domain, but at a deeper level. For combination products, the team must develop a fundamental understanding of the relationship between process parameters, design inputs, and product stability through design characterization. It is important that the product and process are adequately characterized prior to design verification and validation. It is equally important that the test methods and measurement systems are characterized prior to product and process characterization.

Analytical methods are developed to test the identity, potency, and purity of a drug substance or drug product. Analytical methods are also developed to define key characteristics, for example, the particle size distribution of a drug substance or the dissolution rate of a tablet. This is a continuous process that extends over most of the development cycle, but it gets attention during the design domain.

If the combination product is a drug-device or drug-biologic, the level of analytical method validation increases from early to late development as more information becomes available, for example, processes are finalized, and impurities identified and qualified (define). Analytical methods either are generally applicable (e.g., compendia methods described in pharmacopoeias) or are specifically developed for a particular drug substance or drug product. These methods are also developed for excipients (including functional tests) and stabilizing agents, for example, antioxidants or preservatives. Separate microbiological methods are developed to test for microbial contamination, sterility, effectiveness of preservatives, and so on. Cleaning validation requires the development of methods to detect actives in rinses. Typically these test methods are manual, but where possible we recommend that these methods be automated as much as possible. This will not only help to reduce costs for routine testing, but also improve accuracy. In addition, this will help enable transfer of these methods to appropriate laboratories. It is important to remember that acceptance criteria should be established in parallel with method development. Acceptance criteria and the corresponding methods are referred to as specifications, and it is these specifications that must be submitted to health authorities during the development process (e.g., investigational new drag [IND] stage) and in the final new drug application (NDA). Country-specific requirements may lead to different sets of specifications (e.g., drug product release limits of 90–110% for the United States and 95–105% for the EU). In any event, these specifications are defined by guidelines and regulations and need to be justified by data (historical data, results from stability studies, process capabilities, etc.) and safety considerations (e.g., toxicity qualification of impurities).

Process analytical technology (PAT)

This is a framework developed by the FDA for innovative pharmaceutical development, manufacturing, and quality assurance. The framework is founded on process understanding to facilitate innovation and risk-based regulatory decisions by industry and the agency. The framework has two components: (1) a set of scientific principles and tools supporting innovation and (2) a strategy for regulatory implementation that will accommodate innovation. The regulatory implementation strategy includes creation of a PAT team approach to chemistry manufacturing and control (CMC) review and current good manufacturing practice (cGMP) inspections, as well as joint training and certification of PAT review and inspection staff. One of the advantages of PAT is the potential to replace the batch release based on testing of a small number of samples by the parametric release based on data obtained over the entire production cycle. PAT may require the development of different analytical methods (e.g., near-infrared spectroscopy), not typically used for active pharmaceutical ingredient (API) and drug product testing. PAT has been recently introduced and will affect the way test and analytical methods are developed. There are many current and new tools available that enable scientific, risk-managed pharmaceutical development, manufacture, and quality assurance. These tools, when used within a system, can provide effective and efficient means for acquiring information to facilitate process understanding, develop risk mitigation strategies, achieve continuous improvement, and share information and knowledge. In the PAT framework, these tools can be categorized as:

- Multivariate data acquisition and analysis tools
- Modern process analyzers or process analytical chemistry tools
- Process and endpoint monitoring and control tools
- Continuous improvement and knowledge management tools

An appropriate combination of some or all of these tools may be applicable to a single-unit operation, or to an entire manufacturing process and its quality assurance. It must be noted that these tools are no different than DFSS principles and tools. We believe that through the PAT framework, the FDA has given more reasons for the industry to adapt DFSS tools and methods.

Governance

A consistent and sustainable governance process is critical to combination product development. While some companies can go overboard on this, with multiple levels of review, it is important to implement appropriate levels of review during product development. The six-domain process that we recommend in this book expects strong governance/review gates, as a minimum, between two domains. At each review gate a decision as to whether the project should stop, return for further data, or proceed is required after management review. Critical to the success of these reviews are:

- Having appropriate personnel and leadership commitment
- Having all the required facts and documentation for review
- Not treating review as an event with slide presentations only
- Showing commitment to resolve issues identified during review

Process domain

We will focus on process domain in a chapter dedicated to manufacturing of combination products (Chapter 6). However, we want to highlight certain aspects in this domain here.

Technology transfer/scale-up

The purpose of this step is to transfer manufacturing process from development (R&D) to operations (commercial manufacturing). Technology transfer is the process of transferring product development (including packaging), process development, and analytical/test method development knowledge. Typically, full-scale process development work will be performed at the commercial manufacturing site, at the scale planned for launch. The abilities and infrastructure of the receiving site must be constantly considered while development and scale-up are ongoing. It is necessary to lock in the critical materials/API, product, process, packaging, and analytical/measurement systems method information for the final product to be commercialized.

As the full-scale development work is typically performed at the final manufacturing site, it can be performed as part of process transfer or after the transfer from the development site to the manufacturing site. In addition, the analytical/measurement method development report and transfer of analytical test methods to the commercial site or laboratory should be completed prior to technology transfer.

Production scale-up

For combination products, when modifying processes during scale-up, companies need to assess whether these changes will be considered a design change (e.g., a formulation change) and address the impacts thereof. Companies must ensure that these process changes do not invalidate key development results (e.g., clinical trial results, etc.). During production scale-up, companies need to ensure that handling requirements (shipping, storage, handling, etc.) are well understood and appropriately addressed. They also need to meet the requirements for low bioburden material (this requirement should have been identified/addressed earlier in development) since bioburden control during the scale-up activity may require significant effort, facilities, and resources. For combination products that include pharmaceuticals, it is necessary to lock in the critical processing parameters (target and ranges) of the final scale equipment for pharmaceuticals. In addition to this, cleaning validation is extensive for pharmaceuticals. It requires evaluation of the final manufacturing train, the potential impact on cleaning validation of other products, and drug recovery studies on all applicable drug-contacting materials. We highly recommend that the companies ensure that appropriate expertise and approaches are used.

As the full-scale development work for pharmaceuticals is typically performed at the final manufacturing site, it can be performed after or concurrent with the process transfer from the development site to the manufacturing site. The analytical development report and transfer of analytical test methods to the commercial site or lab need to be completed prior to process/technology transfer.

Process development

The commercial process to manufacture the new product in a reproducible manner is developed at this stage. The information and knowledge gained from development studies provide scientific understanding to support the establishing of specifications and manufacturing controls. Key aspects of the product (e.g., drug substance, excipients, etc.) and manufacturing processes that are critical and that present a significant risk to product quality should be identified. Additional studies may enhance the knowledge of product performance and thus establish a "design space" (see ICH guideline Q8) as a basis for "risk-based" regulatory decisions and future manufacturing process changes known as SUPAC (scales-up a post-approval changes).

Process validation

The purpose of process validation is to confirm consistent process performance against the current baseline specifications. A preapproved protocol is required that explains all sampling, in-process monitoring, in-process testing, analytical testing, and acceptance criteria. Each test should be described in detail to ensure full understanding of the rationale for the test and the related acceptance criteria. All development activities should be completed prior to process validation, and the process should be well developed and understood to ensure successful completion of the validation. Also at this stage, inventory for commercial launch is produced.

Supplier selection and qualification

Many combination products are manufactured using components, raw materials, manufacturing materials, and so on, from suppliers (includes service providers, component suppliers, and external manufacturers). As a result, it is the responsibility of the product development teams to screen, assess, and select the suppliers for products and services required to source, manufacture, and distribute a new product. At times these final combination product manufacturers do not have the expertise to deal with suppliers due to the product complexity. For example, if the combination product is a drug-device, the following situation is possible. For any components that make contact with the API, device companies need to have familiarity with pharmaceutical industry practices and regulations. Device companies need to obtain the appropriate knowledge, either from the drug substance supplier or via internal testing, to appropriately utilize the drug substance. The external manufacturers of pharmaceutical constituents have no obligation to provide detailed development information that might be required/desirable for the development of the combination product. In this case, the development team/company can authorize the FDA to access the drug master file (DMF). For externally sourced drug substances, medical device companies should engage the appropriate internal chemical manufacturing and control (CMC) expertise to evaluate the drug substance information provided by the supplier. Device companies will require formulation expertise to convert a drug substance to an appropriate drug product form.

Transfer to operations

The combination product development team should ensure that the appropriate steps have been completed so that the design is correctly translated into production specifications. This step should be performed in parallel with design and development activities. When a pharmaceutical constituent is utilized in the combination product, the product design should be frozen earlier to ensure that the product can be transferred productively and effectively without a need to redo transfer activities. Special attention should be paid to the following during process transfer for the pharmaceutical constituent:

- Need to ensure that specifications set/achieved in lab scale are the same as specifications set/achieved in commercial scale. Ensure that the form of the molecule utilized is the one that has gone through formal technology transfer.
- Pharmaceutical process parameters (e.g., utilities, facilities, environment, etc.) are typically more sensitive than device process parameters.
- Cleaning processes' development and validation for the pharmaceutical constituents are particularly difficult and time consuming.
- Handling requirements for pharmaceutical constituents are typically more stringent.
- Management and control of hold times and storage conditions throughout the manufacturing process are critical.
- Personal protective equipment may be required due to the potency of the pharmaceutical agent.
- Drug Enforcement Agency (DEA)–controlled substances require a site vault and site certification.

Postmarket domain

We will focus on the postmarket domain in a chapter dedicated to postmarket activities for combination products (Chapter 8).

Finally, to wrap up this chapter, we provide an approach that can help product development teams develop products successfully. Table 3.5 is our "equivalency" table that links regulatory requirements such as design control requirements with DFSS. The first column is the flow-down requirements/flow-up capabilities from the DFSS methodology. The second column is the

Table 3.5 Regulatory and DFSS Link

DFSS	Domain	Waterfall model and FDA guidance on design control	Typical DFSS deliverables and actions in the MDI
Define the opportunity[a]	Innovation	Not applicable	Project charter, business case, and project-related risks[b]
Customer wants and needs	Customer[c]	Design and development plan	Project plan (same as design and development plan)
		User needs or potential needs	Plans for obtaining VOC
		Start design history file or equivalent	Obtain VOC—raw and analyzed customer data
		Design review	Define what makes the product special
			Define what makes the potential device hazardous and find out why
			Translate the VOC into high-level systems requirements
			Define the performance requirements for the system requirements
			Initiate requirements management (cascading)
			Initiate project's dictionary (medical specialty jargon)
			Reassessment of medical or clinical risks
		Design input	Measurable high-level systems requirements and performance
		DHF	High-level systems requirements prioritized and with targets or acceptance ranges
		Design review[d]	What would the future user (e.g., customers) evaluate when presented with the final device?
			Medical device report (MDR) analysis (e.g., use the MAUDE database at www.fda.gov)

(*Continued*)

Table 3.5 (*Continued*)

DFSS	Domain	Waterfall model and FDA guidance on design control	Typical DFSS deliverables and actions in the MDI
			Analysis report complaints from similar devices (when applicable)
			Continue requirements cascading
			Review healthcare industry publications such as the Gray and Silver Sheets
			Add definitions to project dictionary
			Research healthcare community journals. Evaluate potential sources of harm such as the ERCI
Medical device functional requirements	Functional	Technical design inputs	Lower-level or tier 2 requirements
		DHF	(Engineering terms) or subsystem requirements
		Design review	Product concepts, prototypes, or computer-modeled devices
		Update design and development plan	Continue requirements cascading
			Add definitions to project dictionary
			Risk analysis such as the systems FMEA
Design and develop medical device, its packaging, and manufacturing process	Design domain and process domain	Approved design inputs	Chosen best concept
		Update design and development plan	Component requirements

(*Continued*)

Table 3.5 (*Continued*)

DFSS	Domain	Waterfall model and FDA guidance on design control	Typical DFSS deliverables and actions in the MDI
		Design output	FMEA/FTA, functional block diagrams
		Risk analysis	Reliability goals and test plan
		Design review	Manufacturing process flowchart
		Design changes	Quality control and assurance plan
		DHF	DfX analysis
			Component and product parameter optimization
			Process parameters and operational ranges (characterization)
			Process optimization
			Add definitions to project dictionary
			Continue requirements cascading
Verify and validate product and manufacturing process	Design and process domain	Design outputs	Validation of test methods
		Design review	Gage repeatability and reproducibility (GR&R), nested DOEs for *in-vitro* diagnostics (IVDs)
		Design verification	Component reliability testing
		Design transfer	Process characterization
		Design changes	Component qualification (including process capability and stability at that level)
		Design transfer	Subassembly/assembly overstress testing
		Risk management	System integration (e.g., Can all the parts work together?)
		Design review	Subassembly process capability and stability

(*Continued*)

Table 3.5 (*Continued*)

DFSS	Domain	Waterfall model and FDA guidance on design control	Typical DFSS deliverables and actions in the MDI
		Design validation	System process capability and stability, including sterilization and packaging
		DHF	Add definitions to project dictionary
			Complete requirements cascading

[a] It is very typical these days to look for alternatives to high-selling drugs or expensive treatments.

[b] Typical risks are, for example, not getting approval from the FDA in a PMA, or not finding an IRB capable or willing to approve a clinical study on an unknown technology. Another risk is the eventual transformation of medical science to a different treatment or drug or biologic that is not even in the picture now.

[c] Not only those who will pay the bills, but keep in mind that MDI is a regulated industry; thus, the FDA, Environmental Protection Agency (EPA), and DEA, among others, are customers as well.

[d] At this early stage of the process, the review is merely aimed at making sure the potential needs or actual needs are understood, and that the process of translating into systems requirements makes sense.

domain in which the NPD process lies. Please note that in any product development process, the domains overlap with each other.

We have tried to establish important activities that need to take place to successfully develop combination products using a domain-based road map. We do acknowledge that there are many more specific activities that could have been added in this chapter, but we felt it important to focus on the key activities rather than all possible activities.

Bibliography

Harry, Mikel and Schroeder, Richard. *Six Sigma, The Breakthrough Management Strategy*. New York: Doubleday, 2000.

Justiniano, Jose and Gopalaswamy, Venky. *Practical Design Control Implementation for Medical Devices*. InterPharm/CRC Press, 2002.

Justiniano, Jose and Gopalaswamy, Venky. *Six Sigma for Medical Device Design*. CRC Press, 2005.

Process Analytical Technology (PAT) initiative: http://www.fda.gov/cder/OPS/PAT.htm (accessed March 2007).

chapter four

Overview of FDA and other regulatory agency expectations

In Chapter 3, we provided the reader with a pathway to develop combination products successfully. One of the key domains in the pathway is the customer domain. While most of the efforts in this domain focus on the customer who is the end user of the product, there are efforts that focus on another "customer," the regulatory agencies such as the Food and Drug Administration (FDA). Chapter 4 discusses meeting the expectations of the FDA and other regulatory agencies, and focuses on the regulatory requirements for combination products in both the EU and the United States. This chapter also provides an overview on the regulatory requirements in Canada, Japan, China, and India at the end of this chapter, as these are other key current or future markets for these products. The regulatory challenges for combination products, which may result in a final product combination of devices, biologics, or drugs, are significant and complex.

There are several critical challenges in developing and bringing to the market novel combination products and meeting FDA and EU regulatory expectations. These challenges, which all development teams must face, are listed below:

1. *PMOA*: An initial challenge the development team would need to face is determining what the primary mode of action (PMOA) of a combination product is going to be. In order to achieve this, the team needs to assess the primary or main intended function of the combination product, and to determine the PMOA, based on this intended use, by analyzing the totality of the function of this combination product. Since FDA responsibilities would depend heavily on the primary function of the combination component, resulting in the main element of the treatment, in device-drug combinations, in this instance, the novelty of the device technology predominates.
2. *Labeling*: Labeling of the combination product also poses several questions for the teams. The approach to cross-labeling should be flexible and discussed not only during labeling negotiations, but also as early as during jurisdiction discussions. If the device labeling is generally

consistent with three key parameters of drug labeling—indication, general mode (route) of delivery, and the drug dosage or dosing schedule equivalence—then the agency should grant Center for Device and Radiological Health (CDRH) jurisdiction for the product, and generally should waive additional clinical showing of drug effectiveness. To accommodate evolution of drugs after marketing, device labeling need only be generally consistent with the labeling of the drug intended to be delivered (e.g., continuous delivery devices for insulin are generally consistent with the drug route of administration). Additionally, even if the device is not consistent in the indication, dose, rate, or route of administration, the differences should be resolved in the device labeling.

In general, the following points can be considered by the company during labeling requirements:

- Label claim for pharmaceuticals/biologics is usually more restrictive and stringent than for devices. While a bench-top/animal model may be adequate for a device claim, clinical data are needed to support label claims for pharmaceuticals/biologics.
- A certified quality person is required for the release of pharmaceuticals/biologics in non-U.S. regions/countries.
- For devices, the sales and marketing force is part of the company structure, compared to the pharmaceuticals/biologics companies, where the sales and marketing function may reside in a different company. This difference of marketing organization adds a level of complexity for device companies not familiar with the separation of the sales and marketing force.
- Additionally, companies need to be aware of reimbursement strategies for the device companies, being different from the pharmaceuticals/biologics companies.
- There is usually a greater amount of scrutiny on reimbursement of the biologics/drugs products in comparison to devices due to the high cost and chronic nature of the product.
- The process for handling adverse drug reports and medical device reports, and how regulatory affairs view them, is very different for devices versus pharmaceutical/biologic products. As a result, the process and requirements for reporting adverse events (AEs) for combination products must be assessed earlier in the planning stage.

Example: labeling and packaging development for a device/biological combination product

If a device is added to a biological product in a combination product, the key differences in labeling that would exist in comparison to developing a device alone would include the following:

- Instruction for use (IFU) is an important component in device labeling.
- Inventory storage control in the hospitals should have requirements for labeling, print size, and colors.
- Due to the typical length of clinical trials, clinical packaging is usually developed without specific details on the commercial requirements. Commercial packaging is oftentimes addressed at a later date. As a result, clinical packaging is typically managed out of R&D, while commercial packaging is handled through the supply chain. The standard operating procedures (SOPs) and processes are different.
- Addition of a biologic would require temperature, humidity studies, and validation of the shipping container, due to the specific requirement to maintain environmental and biologic stability.

3. *Parallel review*: The next challenge involves the requirement for parallel review of two marketing applications for a combination product. The industry needs a clearer understanding and clarification of specific cases where the parallel review path may or may not be required for marketing applications for a combination product. In many cases, a single application should be adequate; however, in certain instances a separate application may be appropriated by the company (a sponsor), especially if the combination product components are expected to have separate distribution or use/reuse patterns, or if the primary jurisdiction for the combination delivery system has been given to a center other than the CDRH, and the delivery device component is capable of being separately defined and reviewed.

Regulatory requirements for FDA/ EU of combination products

Regulation of combination products from an industry perspective poses unique challenges on how these should be reviewed. Additionally, the approaches to regulating combination products are different for the EU and the United States. However, in general, in regards to the regulation of combination products in both the United States and the EU, it is predicated upon the product's primary mode of action (PMOA), although neither market has dedicated combination product regulations.

The regulatory requirements for combination products continue to evolve as industry and agencies are working together to develop specific requirements. There are still no specific regulations or regulatory submissions unique to combination products currently. However, the FDA and industry are working to define the rules in areas such as product labeling and postmarket surveillance; areas such as product life cycle, postapproval changes, how to handle changes to the entire product versus components, changes across multiple companies, new indications, and unified versus separate labeling are still not clearly defined by the agency.

The Office of Combination Products (OCP) was established by the FDA as a result of the Medical Device User Fee and Modernization Act of 2002. The primary function of the OCP is to coordinate responsibilities such as assigning an FDA center to have primary jurisdiction for reviewing a combination product, ensuring timely and effective premarket review of these products by overseeing reviews involving more than one FDA center, ensuring postmarket regulation consistency and appropriateness, serving as a focal point for combination product issues for internal and external stakeholders, and working with FDA centers to develop guidances or regulations to clarify the agency regulation of combination products.

The FDA has established three centers: the Center for Drug Evaluation and Research (CDER), the Center for Biologics Evaluation and Research (CBER), and the Center for Devices and Radiological Health (CDRH). Among these three centers, the 1938 federal *Food, Drug, and Cosmetic Act* (*FD&C Act*) is the common foundation, as drugs, devices, and biologics share many basic features. However, drugs, devices, and biologics each have noticeable differences in their specific requirements for premarket application, labeling, good manufacturing practices (GMP) regulations and postmarket safety reporting, design controls, and other requirements. These differences and specific requirements present an even more complex scenario in regulating combination products. Furthermore, the diversity of combination products and advanced technologies brings on more challenges.

As combination product designs become more complex and their uses more multifunctional, the need for the OCP to assign regulatory jurisdiction to a lead center has become more common. This is a critical process (especially for emerging companies), given that the complexity, risk, time, and cost of developing a product are, in many cases, tied to its regulatory path. To facilitate that process, the FDA issued the *Final Rule on the Definition of Primary Mode of Action of a Combination Product* as a guideline for determining which agency center will have the lead role in regulating a combination product. According to the final rule, PMOA is defined as the "single mode of action of a combination product that provides the most important therapeutic action of the combination product." To make that determination, both industry and the FDA must rely upon, and to some degree are constrained by, the definitions of a drug, device, and biologic as stated in the *FD&C Act*.

The "What are the definitions of a drug, device, or biologic?" Section 201(h) of the *FD&C Act* defines a device as

> an instrument, apparatus, implement, machine, contrivance, implant, *in vitro* reagent, or other similar or related article, including any component, part, or

> accessory, which is ... intended to affect the structure or any function of the body of man or other animals, and which does not achieve its primary intended purposes through chemical action within or on the body of man or other animals and which is not dependent upon being metabolized for the achievement of its primary intended purposes.

As stated, the law implies that the term *chemical action* defines a biologic or drug effect and impacts not only the development of combination under the current system, assignments of combination products to the various centers are made by the OCP via a Request for Designation. Determinations of the device attributes of a noncombination product are made via a 513(g) submission to the Center for Devices and Radiological Health, and for novel devices as well.

A combination product is assigned to one of the agency's three human medical product centers: CBER, CDER, or CDRH. The lead center has oversight responsibility for the review and regulation of the combination product. The lead center often consults or collaborates with other agency components and OCP, as appropriate, to identify and evaluate the information needed for a regulatory submission (e.g., investigational application or marketing authorization).

OCP also works with agency centers to develop guidance and regulations to make the regulation of combination products as clear, consistent, and predictable as possible. Under section 503(g)(1) of the act, a combination product is assigned to a center with primary jurisdiction, or a lead center, based on a determination of the primary mode of action (PMOA) of the combination product. As we mentioned earlier, PMOA is defined as "the single mode of action of a combination product that provides the most important therapeutic action of the combination product." For example, if the PMOA of a device-biologic combination product were attributable to its biological product constituent, the agency component responsible for premarket review of that biological product, that is, CBER, would have primary jurisdiction for the combination product. This approach is simple when the combination product has one PMOA (drug, biologic, or device). However, when more than one component has a PMOA, the decision is complex and may raise issues around the reporting requirements for premarket applications and postmarketing safety.

In streamlining the review of combination products, the FDA established a standard operating procedure (SOP) for the Intercenter Consultative and Collaborative Review Process. This document provides the policies and procedures for FDA staff to follow when requesting, receiving, handling, processing, and tracking formal consultative and collaborative reviews of combination products, devices, drugs, and biologics. The objectives of the SOP are to ensure timely and effective intercenter communication on combination products, as well as the timeliness and consistency of intercenter consultations and collaborations. It is our opinion that this SOP is certainly valuable to the industry, as it can minimize the overall cycle time for product registration or approval.

This guidance describes general information on developmental considerations for products that combine devices, drugs, and biological products and encourages developers to contact OCP for assistance in determining the assignment of a lead center when jurisdiction is unclear or in dispute; the number and type of marketing applications; the premarket review process; and the postmarket regulations, such as adverse event reporting or good manufacturing practice requirements that may be appropriate for a combination product. Figure 4.1 provides an overview of the intercenter collaborative/consultative review process.

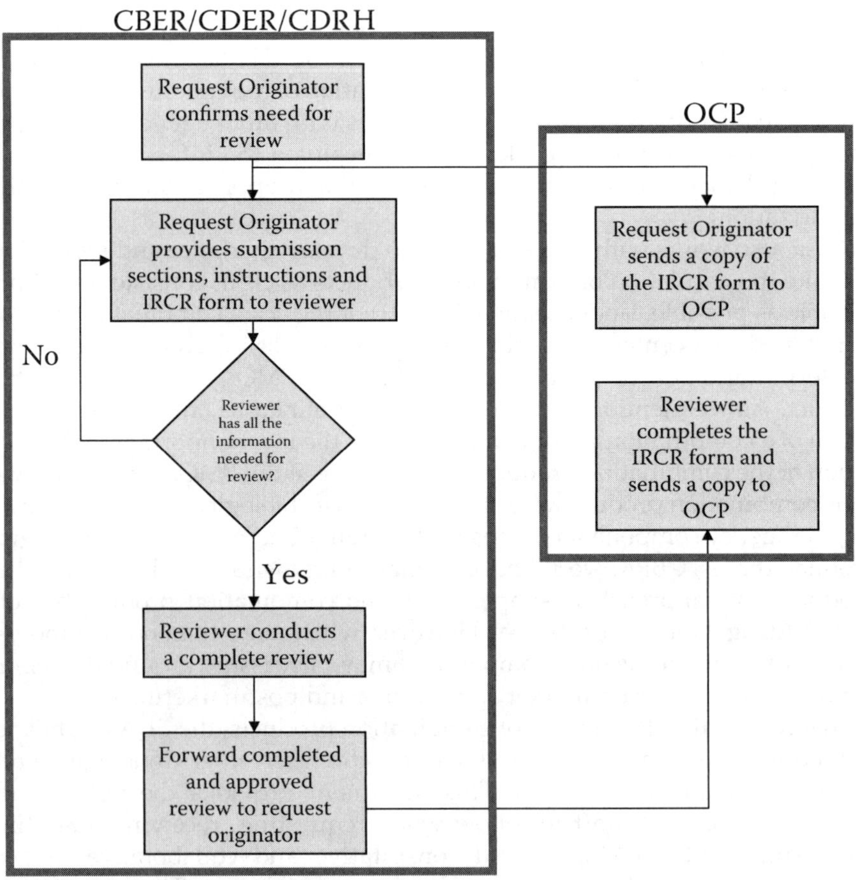

Figure 4.1 FDA's Intercenter review process.

Regulation of combination products by FDA agency

1. ***CDRH*: For products where CDRH is lead, formal meetings are for determination and agreement, and informal meetings can be pre-IDE (investigational device exemption), pre-PMA (premarket approval application), anytime.**
2. ***CDER* or *CBER*: For products where CDER or CBER is lead, pre-IND (investigational new drug), end of phase 2, pre-NDA (new drug application)/BLA (biological license agreement). Table 4.1 shows the product registration by FDA based on the CDRH, CDER or CBER divison.**

Number of marketing applications for a combination product

Another area that the combination product developers deal with is the number of marketing applications needed. How does the FDA address this? We already presented that the OCP assigns a combination product to an individual center (CDER, CBER, or CDRH) that will have primary jurisdiction for its premarket review and regulation, based upon the product's PMOA. It is important to note that a PMOA does not automatically determine the type (510(k), PMA, abbreviated new drug application [ANDA], IND, NDA, or BLA) and number of marketing applications required for the product's

Table 4.1 FDA Product Registration: Approval of Devices, Drugs, and Biologics

	Center for Devices and Radiological Health (CDRH)	Center for Drugs Evaluation and Research (CDER)	Center for Biologics Evaluation and Research (CBER)
Approval to begin clinical evaluation	IDE–Investigational device exemption	IND–Investigational new drug	IND–Investigational new drug
Permission to begin marketing	PMA–Premarket approval application	NDA–New drug application	BLA–Biologic license application
Permission to market a modified product	PMA supplement	NDA or efficacy/ manufacturing supplement (for approved drug)	New license application–efficacy/ manufacturing supplement
Other pathways to marketing	510(k)–Premarket clearance	ANDA–Abbreviated NDA–generic drug bioequivalent to approved drug	N/A

Source: FDA Center for Devices and Radiological Health (CDER).

approval, clearance, or licensure. Depending upon the combination product type, the manufacturer may have to submit a single marketing application for the product as a whole, or two (or more) marketing applications for its constituent parts. It must be mentioned that for most combination products, a single marketing application is sufficient for approval, clearance, or licensure.

When would a manufacturer go for two marketing applications? In some cases, the manufacturer (sponsors) may choose to submit two marketing applications for a combination product because a sponsor would like to receive some benefit that is only possible from a particular marketing application type, for example, in the case of a new drug product exclusivity, orphan drug benefits, or proprietary data protection. The FDA may determine that two marketing applications are necessary, for example, when one constituent part of a combination product is already approved for another use and the approved labeling will need to be changed to reflect its new intended use in the combination product. Let us look at this topic at a deeper level to provide the reader with a better understanding.

When one marketing application may be appropriate

In certain circumstances, a single marketing application may be the only feasible option:

1. Combination products that are chemically, physically, or otherwise bound into a single entity (21 CFR 3.2(e)(1)); drugs and devices generally are approved or cleared only as finished products, not as components for further manufacture.
2. Most copackaged combination products (21 CFR 3.2(e)(2)), particularly those with constituent parts that could not be provided separately, for example, copackaged combination products where one component is not sufficiently finished to support a separate approval/clearance or where the indication exists only in the copackaged configuration.
3. Combination products for which separate applications would create a regulatory inconsistency; the OCP is particularly interested in stakeholder perspectives on these examples and whether there are others for which one application would be the only feasible option.

When two marketing applications may be necessary

From a regulatory perspective, the necessity for more than one marketing application is infrequent; two marketing applications may be needed in the following examples:

1. Combination products where the constituents are separate and complex products.
2. Constituent parts with uses beyond the combination product.

3. To effect labeling revisions for a constituent part already approved for uses that do not include the proposed combination product indication.
4. To apply mechanisms necessary to ensure appropriate regulation, or unique regulatory requirements that are not available under a single marketing application.
5. To maintain regulatory consistency.

Flowcharts

The Office of Combination Products (OCP) provides three flowcharts that display a hypothetical process for determining whether one or multiple marketing applications would be necessary. Figure 4.2 includes our combined version of the three separate flowcharts provided by the FDA.

In conclusion, the following pathways, as indicated in Table 4.2, are normal for products that follow CDRH, CDER, CBER, and combination products regulations.

FDA agency: CDRH (devices)

Examples of the regulatory pathway for several different combination products

Example 1: Drug-eluting stent

In a drug-eluting stent (DES), the stent is coated with or has a drug that is designed to be released into the surrounding area in a time-controlled manner. The reason for the time-release process of the drug is to reduce restenosis by slowing down cell growth. As we presented in Chapter 2, the FDA approved the Cordis CYPHER Coronary Sirolimus–Eluting Stent in April 2003. In this case, the PMOA is the medical device, that is, the stent as it opens the artery. The secondary action is by the drug, which prevents inflammation and restenosis of the artery. This combination product is therefore regulated by CDRH under device provisions as a PMA.

Detailed regulatory pathway for the DES in the United States

This combination product is composed of three components: the stent platform and delivery system, which have fallen to the CDRH for review; the polymeric carrier in which a drug is loaded (polymeric coatings have also traditionally been reviewed by the CDRH); and the drug substance, which has fallen under the CDER branch of the FDA. The combination product as a whole has been reviewed by both centers in the United States.

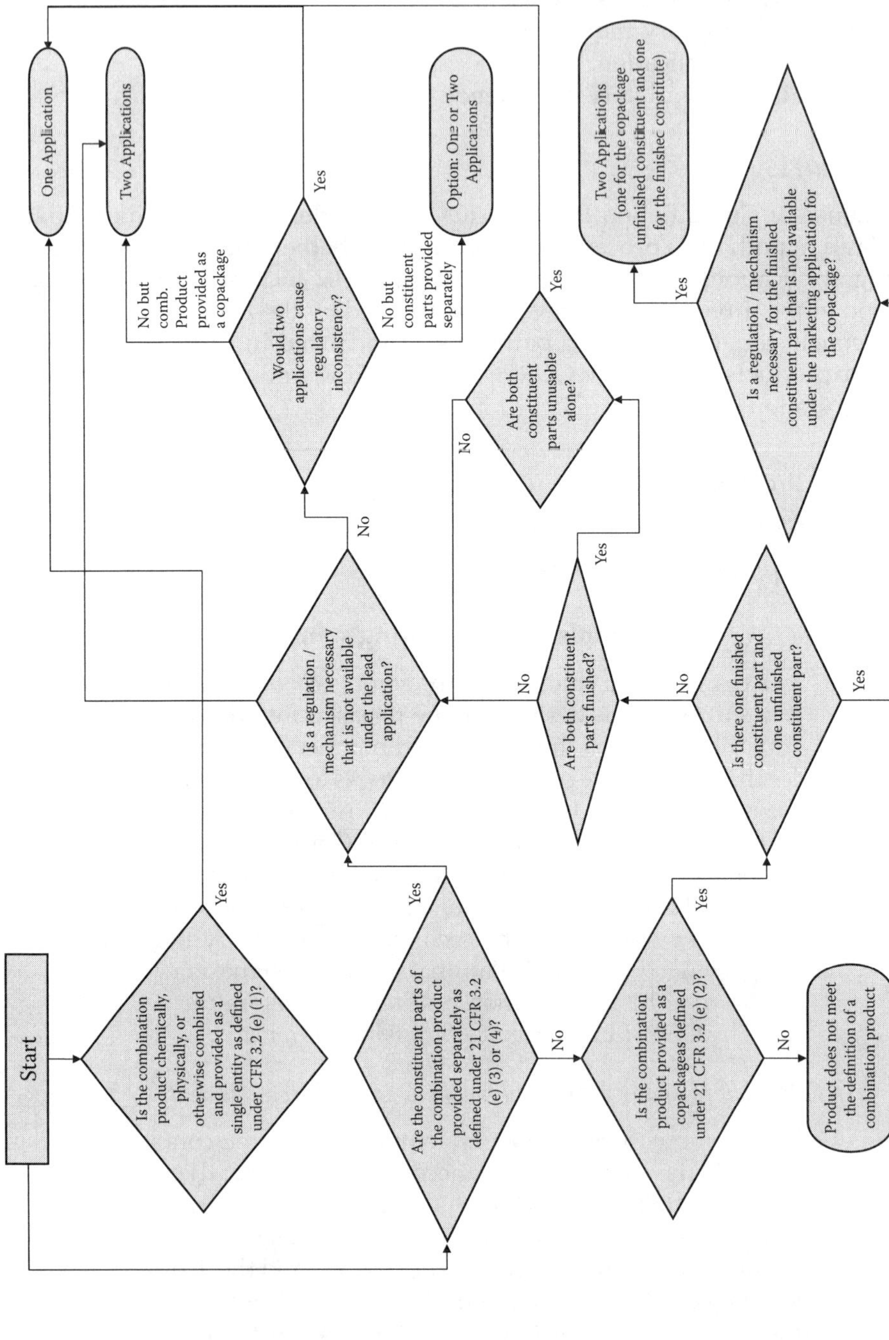

Figure 4.2 Process for determining the number of marketing applications necessary.

Table 4.2 EU consultation Dossier's Requirements

Item no.	Required information
B.3 (a)	General information—Medical device description, including manufacturer's claim on inclusion of medicinal substance and risk analysis.
B.3 (b)	Qualitative and quantitative particulars of the constituents—Medicinal substance description and the quantity in each medical device. Details on substance if modified.
B.3 (c)	Description on method of manufacture—Focus on incorporation of the medicinal substance in the device.
B.3 (d)	Starting materials controls—Medicinal substance specification, European Pharmacopoeia reference, or a drug master file reference if available.
B.3 (e)	Control tests carried out at intermediate stages of the manufacturing process of the medical device if directly relevant to the quality of the medicinal substance.
B.3 (f)	Finished product control tests—Qualitative and quantitative tests carried out to control the medicinal substance in the device.
B.3 (g)	Stability—Medicinal substance stability information throughout the shelf-life of the device.
B.3 (h)	Toxicity—Reference to the substance's known toxicological profile. With new active substances, studies on toxicity and biocompatibility are needed.
B.3 (i)	Reproductive function (see B.3 (h)).
B.3 (j)	Embryo/fetal and perinatal toxicity (see B.3 (h)).
B.3 (k)	Mutagenic potential (apply B.3 (h)).
B.3 (l)	Carcinogenic potential (apply B.3 (h)).
B.3 (m)	Pharmacodynamics—Intended action of the substance in the context of use in the medical device.
B.3 (n)	Pharmacokinetics—Not required in the majority of cases. Describe substance's local and system's availability.
B.3 (o)	Local tolerance—Address this particularly if the medical device's route of administration is different than at the normal.
B.3 (p)	Clinical documentation.
B.3 (q)	Labeling.

Source: Adapted from MEDDEV 2.1/3 rev. 2 (July 2001).

The regulations governing combination products are set forth in 21 CFR Part 3.

A request for product jurisdiction for the DES was submitted early in the process with the lead center designated as CDRH, and CDER serving in a consultative role. Other divisions with potential involvement would be Division of Cardiovascular Devices, Division of Mechanics and Materials Science (CDRH), Office of New Drug Chemistry (CDER), and so forth.

Marketing submission

The appropriate marketing submission route for the DES is a premarket approval application (PMA) with CDRH as the lead branch of FDA. The appropriate application for investigation of these stents in a U.S. clinical trial is an investigational device exemption (IDE). The manufacturing inspections for the DES were conducted by CDRH, the lead center, with strong involvement from CDER's Office of New Drug Chemistry. The FDA reviewed application of the drug regulations (current good manufacturing practices [cGMP]) for the manufacturing of the drug substance, and the device quality system regulations (QSR) for the finished product.

Preclinical testing

For preclinical testing, the finished sterilized product characterization is key for the studies. Although changes and design improvement can occur during a product development process, the manufacturer has to be able to characterize the actual product intended for a study for the FDA to assess the results of the clinical trial and how these can apply to the commercially distributed product.

Certain key areas include characterization of the coating of the drug substance, *in vivo* and *in vitro* elution test methods, specifications to characterize the drug release rate, and initial data to support stability of the drug and the polymeric carrier. Additionally, animal studies are also needed to assess safety prior to conducting human clinical trials.

Clinical evaluation

The FDA looks for reasonable assurance of safety and effectiveness, and therefore, the clinical trial design should address and meet both of these objectives. A randomized controlled clinical trial needs to be performed for study endpoints. The primary endpoints should include at least one clinically meaningful endpoint. For a robust DES trial, use of independent core labs, clinical event committees, and an online, active data safety monitoring board is critical for timely evaluation of the results and the monitoring of the clinical trial performance.

Example 2: Spinal fusion device coated with therapeutic protein (CDRH)

This product is intended to treat degenerative disc disease. In this case, there are two modes of action:

1. Mechanically maintain intervertebral spacing (device component).
2. Encourage formation of bone within fusion cage.

The PMOA in this case is attributable to the device component's action to mechanically maintain spacing and stabilize the spine. Spinal fusion can be obtained without protein.

Since the protein cannot maintain space and stabilize alone, it has a secondary action.

Example 3: Drug-eluting disc: drugs (CDER)

The primary mode of action of the drug-eluting disc is cancer chemotherapy of brain tumor, and the secondary actions are local drug delivery of drug by device. As a result, this product was regulated by the CDER under drug provision.

Example 4: Vitagel™ Surgical Hemostat: biologics (CBER)

The Vitagel Surgical Hemostat is a product that assists the body in clotting blood. Vitagel includes the biological component thrombin, an enzyme that assists in the clotting of blood. The primary mode of clotting (PMOA) for Vitagel is the formation of a protein (collagen/fibrin) clot that serves as a physical barrier to blood flow. Vitagel is intended to assist in clotting when conventional means fail or are impractical. This product was regulated by the CBER.

The four examples above help the reader to understand the regulatory pathway for combination products, albeit at a high level. One can clearly see how PMOA determination is key in helping combination product development teams understand the regulatory pathway. Interaction with the FDA early and often can help a company in PMOA determination and beyond.

FDA interactions: early interaction and communication with the FDA

The FDA strongly encourages early communication and discussion among developers, FDA review components, and, as appropriate, the OCP. Early dialogue allows developers to obtain initial feedback on the kinds of preclinical and clinical testing that may be necessary. Such communication may identify critical issues for product development and help to ensure an efficient development and approval process. Furthermore, early and frequent communication provides the opportunity for the FDA to establish its intercenter review team and to develop the appropriate scientific expertise to facilitate timely and efficient reviews of any future submissions.

The CBER, CDER, and CDRH provide guidance on milestone/collaboration meetings throughout the development process and submission of investigational and marketing applications. Preinvestigational (pre-IND and pre-IDE) meetings are particularly useful for discussing innovative combination products. Premarketing application meetings are also helpful to discuss application content, as well as the sequence and timing of modular applications, or when more than one marketing application will be submitted for the combination product.

The lead center should be contacted to schedule meetings in accordance with the procedures and milestones applicable to the lead center. We encourage developers to request participation from relevant review components from both the lead and consulting centers, where appropriate. In addition, the OCP is available formally or informally to address jurisdictional, developmental, premarket review, and postmarket concerns.

Combination product regulation in the EU

There are no specific regulations or regulatory body with jurisdiction for combination products in the EU. Instead, the regulation of combination products is implied within the Medical Devices Directive, which states that the regulation of the combination product will depend on its primary mode of action, which is similar to that in the United States. Additionally, combination product review occurs via a consultation between a notified body and a competent (health) authority products industry. Combination products are now recognized as one of the fastest-growing sectors in the overall medical sector. There is no official differentiated status for combination products; additionally, there also is no system that has been formalized, similar to the U.S. Request for Designation (RFD), in Europe.

The first step in determining which European regulatory requirements are applicable for a combination product is to determine the *product designation*. The appropriate regulatory pathway will be driven by the correct product designation. Classifications for traditional pharmaceutical drugs or medical devices are easy to identify. Classification of a combination product, however, such as a drug-device or a biologic-device, would require a thorough review of the regulations of the particular nation in which the product is going to be marketed.

The current EU *Medical Devices Directive* (*MDD*) (93/42/EEC) describes a medical device, in part, as a product that "does not achieve its principal intended action in or on the human body by pharmacological, immunological or metabolic means, but which may be assisted in its function by such means." As noted, specific combination product regulations do not currently exist in the EU, but the *MDD* lays out the regulatory pathway for combination products that have a medical device as a component. *MDD* clause 3 specifies that devices intended to administer a medicinal product (as defined in the *Medicinal Products Directive* [*MPD*], 65/65/EEC) are regulated as medical devices and governed by the *MDD*. However, if that device and the medicinal product are "a single integral product which is intended exclusively for use in the given combination," then the product is regulated as a drug and is governed by the *MPD*. Finally, clause 4 of the *MDD* states that if "a device incorporates, as an integral part, a substance which, if used separately, may be considered to be a medicinal product, and which is liable to act upon the body with action ancillary to that of the device," that device is governed by the *MDD*.

Article 1, section 3 of the MDD also places drug delivery products as medical devices subject to the MDD when they are two separate components: where a device is intended to administer a medicinal product within the meaning of article 1 of Directive 65/65/EEC, that device shall be governed by the present directive (Medical Devices Directive, 93/42/EEC), without prejudice to the provisions of Directive 65/65/EEC with regard to the medicinal product.

Mode of action

The mode of action is defined as "the means by which a product achieves its intended therapeutic effect or action. For purposes of this definition, 'therapeutic' action or effect includes any effect or action of the combination product intended to diagnose, cure, mitigate, treat, or prevent disease, or affect the structure or any function of the body." Products may have a drug, biological product, or device mode of action. Because combination products are comprised of more than one type of regulated article (biological product, device, or drug), and each constituent part contributes a biological product, device, or drug mode of action, combination products will typically have more than one mode of action.

In general, regulation of combination products in both the United States and the EU is predicated upon the product's PMOA, although neither market has dedicated combination product regulations. PMOA asks the question: Which element of the product, the drug, device, or biologic, produces the clinical effect for the patient? We gave the example of the drug-eluting stent earlier. Another example is a prefilled inhaler containing an aerosol drug, which is a drug-device combination product where the primary mode of action is affected by the drug. Note that the inhaler device is the drug delivery system and does not by itself produce a clinical benefit for the patient. In this case, the inhaler device is just a delivery system with little to no clinical benefit to the patient, with the main clinical effect obtained by the drug.

Another common type of combination products is a medical device that incorporates a drug as an ancillary medicinal substance. In this case, the product's primary mode of action is affected by the medical device's mechanical function, with the drug providing secondary effects. One prominent example of this type of combination product is the drug-eluting stent: the primary mode of action is mechanical (the stent's struts hold the artery open), and the drug is present to prevent excessive scar tissue from forming on the stent and potentially causing a new obstruction. Here, the drug's action is secondary to the device.

The current EU *Medical Devices Directive* (93/42/EEC) describes a medical device, in part, as a product that "does not achieve its principal intended action in or on the human body by pharmacological, immunological or metabolic means, but which may be assisted in its function by such means." As noted, specific combination product regulations do not currently exist in the EU; however, the *MDD* lays out the regulatory pathway for combination products that have a medical device as a component. *MDD* clause 3 specifies that devices intended to administer a medicinal product (as defined in the *MPD*, 65/65/EEC) are regulated as medical devices and governed by the *MDD*. However, if that device and the medicinal product are "a single integral product which is intended exclusively for use in the given combination," then the product is regulated as a drug and is governed by the *Medicinal Products Directive*.

Finally, clause 4 of the *MDD* states that if "a device incorporates, as an integral part, a substance which, if used separately, may be considered to be a medicinal product, and which is liable to act upon the body with action ancillary to that of the device," that device is governed by the *MDD*.

The European Commission (EC) provides further guidance on the demarcation between medical devices and medicinal products in MEDDEV 2.1/3 rev. 2 (July 2001). This document includes a comprehensive and helpful list of example products in various categories to assist notified bodies (NBs) and manufacturers in identifying the appropriate regulatory pathway for a combination product.

How are the regulations applied to combination products?

Medical devices that administer a drug but are not integral to the drug (e.g., an infusion pump) are regulated as medical devices under the *MDD*. Medical devices that are integral to the drug being administered, such as a drug inhaler, are regulated as drugs under the *MPD*. Medical devices that incorporate a drug as an ancillary medicinal substance, where the medical device has the primary mode of action (for example, bone cement containing an antibiotic), are regulated as medical devices under the *MDD*. It is important to note, however, that this latter product category has requirements above and beyond the *MDD*, as discussed below. During discussion of the proposed *MDD* revisions currently under consideration in the EU, the EC has indicated that combination products with no predominant mode of action (that is, where the drug and device are equally important to the clinical effect) will be regulated under the *MPD*.

The *MDD* carries through the theme of ancillary medicinal substances in clause 7.4 of the essential requirements (Annex I), which states in part:

> *Where a device incorporates, as an integral part, a substance which, if used separately, may be considered to be a medicinal product as defined in Article 1 of Directive 65/65/EEC and which is liable to act upon the body with action ancillary to that of the device, the safety, quality and usefulness of the substance must be verified, taking account of the intended purpose of the device, by analogy with the appropriate methods specified in Directive 75/318/EEC.*

The consultation procedure, MEDDEV 2.1/3 rev. 2 (July 2001), outlines the procedures for fulfilling this *MDD* requirement for medical devices containing an ancillary medicinal substance. Since notified bodies are authorized to review only medical devices, MEDDEV 2.1/3 requires that a competent authority (CA) be consulted in reviewing a medical device containing an ancillary medicinal substance. This is known as the consultation procedure. In this procedure, the notified bodies consult an EU member state CA to assess the "safety, quality and usefulness of the substance" as a component of the medical device in question. This procedure is intended to determine whether the medicinal substance is suitable for its intended use in the medical device, and whether the risks of using the medicinal substance are justified by its benefits.

It is important to note that the consultation procedure is between the NB and the CA, and not between the manufacturer and the competent authority. Therefore, it is incumbent upon the NB to determine the need for a consultation and to arrange one with a CA. However, it must be noted that consultation fees due to the CA are the manufacturer's responsibility.

The CA is expected to review the medicinal component "by analogy to" Directive 75/318/EEC, which specifies the required medicinal product marketing authorization documentation for CA review. Section B.3 of the MEDDEV document lists the information the dossier should include for the consultation. These data are outlined in Table 4.2.

Medical device manufacturers developing their first combination product with a drug must be prepared to present an extensive drug component dossier for CA review. As mentioned previously, the MEDDEV guidance lists the data required in this dossier. CAs may request whatever documentation they believe necessary to judge the drug's utility and safety for its intended purpose in the medical device. The CA selected by the author's firm requested that a dossier be submitted in Common Technical Document (CTD) format, specifically Module 3—Quality. CTD is an international format for pharmaceutical marketing applications that defines the submission's structure and content. It requires highly detailed information on the specifications, manufacturing, quality, and test methods of the drug substance (active ingredient), excipients, and finished drug product.

Much of the information a consultation dossier requires can be addressed relatively easily if the manufacturer has an EDQM (European Directorate for the Quality of Medicines) Certificate of Suitability (CoS) for the drug. A CoS guarantees that the drug complies with its particular European Pharmacopoeia monograph, and therefore is of assured chemical purity and microbiological quality. The combination of the EDQM certificate and European Pharmacopoeia contains much of the drug assessment information a CA requires for a drug-device combination consultation procedure. Alternatively, or in addition to the CoS, the drug supplier may have a Drug Master File lodged with one or more CAs.

In the absence of a CoS or Drug Master File, the device manufacturer has to rely on the drug manufacturer or supplier to provide the necessary consultation dossier information. This is a difficult proposition if the two parties have no contractual relationship, or the device manufacturer is purchasing such small quantities that the drug manufacturer is not inclined to assist in the consultation procedure. It is therefore critical that the availability of the CoS, Drug Master File, and other documentation be considered very early in the product development process, when potential drug suppliers are being evaluated. A supply agreement between the device manufacturer and the drug supplier, including guaranteed access to required information, is imperative if no CoS or Drug Master File exists.

Assuming that the data are available, the NB schedules the review with the competent authority and forwards the dossier. The review process may be lengthy and the CA might generate multiple rounds of questions that must be addressed. The initial review period and follow-up can last between 3 and 6 months, with the whole CA process taking about a year.

At the consultation's conclusion, the competent authority issues an opinion to the NB about whether the medicinal substance's proposed use in the medical device is safe, meets quality standards, and is useful in the proposed context. The NB factors this opinion into its decision regarding the device's CE (Conformité Europeénne) marking. The NB is expected to inform the CA of its final decision. If the NB, despite a negative opinion from the CA, wishes to issue a CE certificate, it is expected to consult the CA prior to doing so. One final comment: The addition of an ancillary medicinal substance to a medical device significantly prolongs and complicates the CE marking process. Understanding the documentation requirements and consultation process, and communicating this information within project teams early in a combination product's development are critical to success in receiving regulatory approval.

Sponsors of combination products are also presented with complex issues such as those related to determining primary mode of action and handling adverse event reporting, labeling, and postapproval changes. In many cases, many drug delivery devices are developed by device companies for uses with approved drugs or biologics. There are additional challenges if the device allows the drug or biologic to be used for new and different indications for use, or there is a different mode of delivery, for example, an IV versus a topical application.

The regulatory pathway for a drug-eluting stent in the EU

An example of the regulatory pathway for a drug-eluting stent (DES) in the EU is shown below to highlight the similarities and differences between the EU and the United States for the same product.

The regulatory pathway for a DES in Europe is as follows, starting with the components of a DES.

Component 1: The bare-metal stent

The questions around the bare stent are whether the stent is manufactured from known and well-characterized materials already in use. This can simplify the development of the clinical and regulatory strategy since there would potentially be a regulatory approval and medical history on the product.

A second issue is to see if the bare stent has been approved for sale in Europe previously. If the bare stent has a CE mark and the company is going to use this bare stent, this can provide a robust foundation for assimilating the DES design dossier as well as contribute to the clinical studies strategy. A comparison of the DES versus other bare-metal stents in reference to characteristics such as radial strength, opacity, vessel wall coverage, flexibility, deliverability, and so on, needs to be performed. A strong history can provide a foundation to overall claims and compliance to relevant standards.

Component 2: The drug (active ingredient)

A few questions need to be addressed by the development team to address the pathway for the drug component as follows:

- Is the stent manufacturer also the developer/manufacturer of the drug? If yes, the product development team and regulatory would have substantial data. These data can be used to support the use of the drug in its new application.
- Another key question is what is the principle mode of action (PMOA) of the drug? As noted, the drug is ancillary to the therapeutic effects of the device for the DES, for the product to fall under the jurisdiction of

the MDD. However, the actual mode of action of the drug is critical in making plans for the preclinical testing, clinical studies, and so on.

- What animal studies have been performed with the drug on its own? This information can be used in appropriate design of preclinical animal studies for the combined product.
- Is the drug approved in Europe, the United States, or any other regulated market? For an approved drug, there may be a possibility to cross-reference existing marketing approvals or other regulatory files. This would reduce the amount of work and evidence the manufacturer needs to generate and supply to the regulatory agencies.

Component 3: Drug delivery (carrier)

Some of the questions regarding the drug delivery carrier are as follows:

- How is the drug going to be retained on the stent? If the drug can only be fixed onto a stent by means of a polymer or carrier, is this material well known and characterized? The carrier coating placed on the stent will have a critical role in the overall product design. It will regulate the release kinetics of the drug, impact the biocompatibility of the device, and may react with tissue in an antagonistic way.
- Has the proposed material been used in cardiovascular applications before? If yes, any device that has utilized the carrier material in the past may provide valuable insight into its suitability for DES applications as well as provide guidance on preclinical testing.
- Do data exist for the stent-plus-carrier combination? If yes, the manufacturer can save a great deal of preclinical testing, bench work, and clinical design issues.

Clinical studies

In reference to clinical studies for the DES, several are likely to be required, ranging from first-in-man studies to large efficacy trials. Careful planning, review of current data, and awareness of the European/national regulations will all be key factors in meeting the relevant requirements and ensuring timely introduction of the product to the market.

Investigation brochure

During commencement of clinical trials in the development process, the regulatory affairs (RA) representative has to put in place relevant documents for submission to the ethics committees and medical device competent authorities. Key questions addressed are as follows:

- Has the device been adequately described with suitable diagrams?
- Has interaction of the drug component with the stent and carrier been suitably explained?

- Have animal study data been compiled and included appropriately?
- Does the investigator's brochure meet the requirements of the international clinical study standard?

Notified body selection

It is key that the notified body (NB) has experience with both cardiovascular implants and medical devices incorporating a medicinal substance (combination products). This would be critical in ensuring that the manufacturer does not waste time educating the NB. It is also key to note with which medicinal substance competent authority (MSCA) the NB works. This is an important aspect of the final regulatory submissions. Does the MSCA used by the NB have experience in the cardiovascular device arena? If yes, the MSCA that understands the concepts behind the therapeutic application of the device is more likely to be able to make a reasonable assessment of the medicinal content of the product.

Design dossier

The design dossier should contain all the necessary contents for submission to a notified body. An adequate design history has to be compiled for both the metal stent and the combination product with the pharmaceutical component. Major subcontractors in regards to sterilization and the medicinal component need to be documented. An adequate risk analysis of the finished DES has to have been performed and included.

The information for *drug-related contents* required is as follows:

- Qualitative/quantitative particulars on the constituents, chemical description of the drug, and the amount incorporated into the device.
- Description of the manufacturing processes—Specifically related to the incorporation of the drug rather than the overall device or drug manufacture.
- Control of starting materials—The specification for the medicinal substance. A drug file may also be required based on the novelty/application of the drug.
- Stability—Information to establish that the medicinal product maintains its desired function for the lifetime of the device, when stored in the recommended manner, poststerilization.
- Toxicity—If the drug has a known toxicology profile, this may be referred to; if not, toxicity tests should be supplied, which should be a standard pharmaceutical toxicity test, but may include toxicity and biocompatibility testing of the overall device (data on toxicity for reproduction function, fetal, perinatal, mutagenic, or carcinogenic potential).
- Pharmacodynamics—Details of the intended action of the drug in the device context.

- Pharmacokinetics—Studies specifically relating to the release of a new active substance, or a medicinal substance used for a new indication from the device, the subsequent distribution, and elimination.
- Local tolerance—The route of exposure to the medicinal substance may be different from its conventional application.
- Clinical documentation—Usefulness of the medicinal substance in the medical device should be addressed by clinical data in existence.

If a medical device incorporates a medicinal substance whose action is ancillary to the medical device function, the combination product is regulated as a class III medical device under the European Medical Devices Directive (MDD).

Global regulatory requirements

For the most part, the regulatory requirements of countries other than the United States and the EU are attempting to align with FDA or EU requirements. In the next few sections we have attempted to provide an overview of the basic regulatory requirements in countries such as Canada, Japan, China, and India. These countries either have had or will have strong demand for combination products due to established (Canada and Japan) or growing (China and India) economies. We will first discuss the general regulatory approach in these countries and then, to the extent possible, provide approaches for combination product approval.

Canada

In Canada, Health Canada, a government department equivalent to the U.S. Department of Health and Human Services, is responsible for the regulation of drugs, medical devices, natural health products (dietary supplements), and certain medical aspects of foods. As shown in Figure 4.3, Health Canada is headed by the Minister of Health and the Health Products and Food Branch (HPFB), which includes several directorates: the Biologics and Genetic Therapies Directorate, the Marketed Health Products Directorate, the Therapeutic Products Directorate, and the Food Directorate. The key regulations are the *Food and Drug Regulations*, the *Medical Devices Regulations*, and the *Natural Health Products Regulations*.

The HPFB's mandate is to take an integrated approach to managing the health-related risks and benefits of health products and food. The HPFB evaluates and monitors the safety, quality, and effectiveness of the thousands of drugs, vaccines, and medical devices.

The HPFB maintains postmarket safety by monitoring all health products available for sale in Canada for compliance with manufacturing, advertising, and labeling regulations and guidelines, and monitors expected and unexpected health risks such as adverse reactions to drugs. HPFB also works

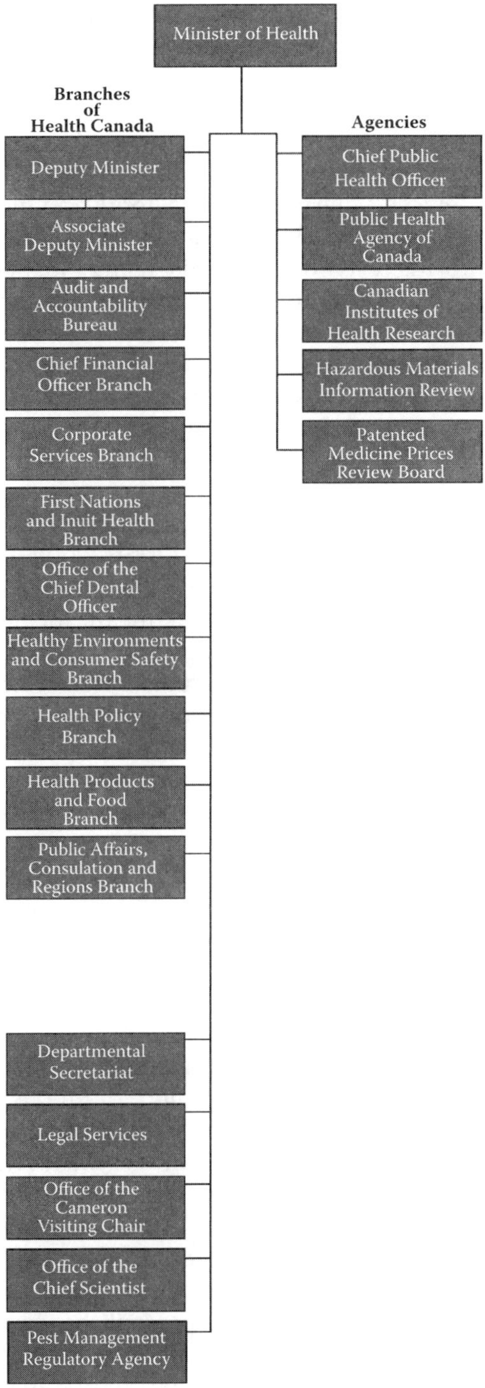

Figure 4.3 Branches and agencies in the Ministry of Health in Canada.

with international partners to harmonize global approaches to biotechnology regulation and use the best-available scientific research and knowledge to evaluate biotechnology-derived products. Within HPFB, postmarket surveillance activities are the responsibility of the Marketed Health Products Directorate (MHPD).

Products that HPFB regulates include pharmaceuticals, natural health products, medical devices, veterinary drugs, food, and biologics. The HPFB Inspectorate is responsible for overseeing the compliance of all regulated products by monitoring establishment licensing, inspections, investigations, and laboratory analysis. To obtain licenses for any HPFB-regulated activity and to renew licenses each year, companies need to demonstrate to the branch that all procedures and processes adhere to appropriate standards. Internationally recognized good manufacturing practices must be followed at Canadian sites involved in the production of drug products. These sites are thoroughly inspected on a regular basis by highly trained professionals.

Pharmaceuticals: Prescription and nonprescription drugs

The Health Products and Food Branch of Health Canada is the federal authority that regulates all pharmaceuticals meant for human use in Canada. Pharmaceuticals are mostly synthetic products made from chemicals for therapeutic use. HPFB's Therapeutic Products Directorate (TPD) is responsible for the regulation and evaluation of prescription and nonprescription pharmaceuticals in Canada. All pharmaceuticals for use by humans in Canada are subject to the *Food and Drugs Act* and its regulations.

New pharmaceuticals are carefully reviewed by TPD before being authorized for sale in Canada. Pharmaceutical manufacturers must submit substantive scientific evidence of a product's safety, efficacy, and quality, which the scientists at HPFB review to determine whether potential risks from the new pharmaceuticals are acceptable when balanced against the positive effects of the drug.

Medical devices

Health Canada, through its Health Products and Food Branch, is responsible for medical devices in Canada. HPFB monitors and evaluates the safety, efficacy, and quality of diagnostic and therapeutic medical devices, so that consumers and healthcare professionals can use them with confidence. All medical devices in Canada are subject to the Food and Drugs Act and its regulations.

Medical devices are categorized into four classes based on the level of risk associated with their use. Class I devices present the lowest potential risk (e.g., thermometers), and class IV devices present the greatest potential risk (e.g., pacemakers). Class II, III, and IV devices receive increasingly rigorous reviews, and have to be licensed before being sold in Canada. Class I devices do not require licenses, but manufacturers must ensure that devices are designed and manufactured to be safe, as required by the Medical Devices Regulations.

Medical Devices Bureau

The Medical Devices Bureau, part of the Therapeutic Products Directorate in HPFB, does this work of reviewing class III and IV medical devices, whether they are manufactured in Canada or abroad, before being authorized for sale in Canada.

The Medical Devices Bureau also works closely with the HPFB Inspectorate to monitor approved medical devices for compliance with Canadian Medical Devices Regulations and Health Canada guidelines, after being available on the Canadian market. If a medical device's safety or its effectiveness is questionable, the manufacturer may be required to recall or refit the device. If necessary, the bureau will suspend a product's license.

Combination products

In reference to combination products, there is a drug–medical device combination products policy in Canada, whose purpose is to ensure timely access to drug–medical device combination products by establishing a single-window approach and more efficient submission processing system, while at the same time ensuring that combination products marketed in Canada are safe, effective, and of high quality.

Currently, sponsors of drug–medical device combination products must satisfy the requirements of two sets of regulations. The drug component of a combination product must comply with the *Food and Drug Regulations,* and the device component must comply with the *Medical Devices Regulations.*

The first reaction, to the statement above, from a combination product manufacturer (sponsor) can vary from surprise to frustration. The Therapeutic Products Directorate and the Biologics Genetic Therapies Directorate (BGTD) are aware of the regulatory burden that this creates for sponsors and the disincentive it presents to marketing combination products in Canada. The directorates believe that the risks associated with a combination product can be managed appropriately under one set of regulations. This approach would harmonize regulatory requirements with both the United States and the European Union and would assist in the development of mutual recognition agreements with those jurisdictions.

With this policy, drug-device combination product classification decisions will consider the PMOA by which the claimed effect or purpose of the product is achieved. The entire product will then be regulated under either the *Food and Drug Regulations* or the *Medical Devices Regulations.*

Ultimately, it will be necessary to amend the *Food and Drugs Act,* the *Food and Drug Regulations,* or the *Medical Devices Regulations,* or all, to provide an appropriate regulatory framework for new and emerging therapeutic products that are difficult to define under current frameworks, including combination products.

Classification of drug–medical device combination products by therapeutic products committee: decisions

The following combination products were classified by the Therapeutic Products Classification Committee in accordance with the *Policy on Drug–Medical Device Combination Products*, October 20, 1998. This also included products that are not combination products but where the classification of either drug or device was difficult to determine. The program had the following policy:

- A combination product will be subject to either the *Medical Devices Regulations* or the *Food and Drug Regulations* according to the principal mechanism of action by which the claimed effect or purpose is achieved.
- Where the principal mechanism of action by which the claimed effect or purpose is achieved by pharmacological, immunological, or metabolic means, the combination product will be subject to the *Food and Drug Regulations,* unless that action occurs *in vitro,* without reintroducing a modified cellular substance to the patient, in which case the product will be subject to the *Medical Devices Regulations.*
- Where the principal mechanism of action by which the claimed effect or purpose is not achieved by pharmacological, immunological, or metabolic means, but may be assisted in that effect or purpose by pharmacological, immunological, or metabolic means, the combination product will be subject to the Medical Devices Regulations.

This listing is not all-encompassing, due in part to the complexity and uniqueness of combination products; therefore, it should be used only as a guide in determining the status of combination products.

Classification examples of products

1. Combination products that have been classified as drugs:
   - Prefilled syringes
   - Patches for transdermal drug delivery
   - Implants whose primary purpose is to release a drug
   - Wound dressings whose primary purpose is to deliver a drug
   - Dental products impregnated with a drug whose primary purpose is to deliver a drug
   - Red blood cell processing solutions
   - Contrast media
   - Peritoneal dialysis solutions
   - Alcohol swabs
2. Combination products that have been classified as devices:
   - Drug-coated devices such as catheters, shunt sensors, or pacemaker leads
   - Drug-impregnated devices

- Wound dressings and surgical barriers containing an antimicrobial agent
- Wound dressings whose primary purpose is to act as a barrier to pathogens
- Blood bags containing anticoagulant or preservation solutions
- Bone cement–containing antibiotic and novel bone void fillers, for example, collagen matrix with bone morphogenic protein
- Injectable collagen
- Sodium hyaluronate nasal solution
- Urea breath test (accessory to device)
- Device for ex vivo photodynamic cell processing

3. Combinations of drugs and devices to which this policy does not apply and which must comply with both the Food and Drug Regulations and the Medical Devices Regulations:
   - Kits (e.g., epidural tray containing drugs and devices, first aid kit containing drugs and devices)
4. Products for which neither set of regulations applies:
   - Organ preservation solutions
   - Minimally manipulated tissue

Japan

The Japanese Ministry of Health, Labor, and Welfare (MHLW), the regulatory authority in Japan, has made significant strides in harmonizing its medical device regulatory practice with international standards. Examples include the establishment of the Pharmaceutical and Medical Device Agency (PMDA) in the year 2004 and implementations of the revised Pharmaceutical Affairs Law (PAL) in the year 2006. Table 4.3 provides details on how this ministry is organized. The 2006 implementation of the PAL revisions has demonstrated Japan's objective of aligning its regulatory practices with other countries. The PAL ensures the safety, efficacy, and quality of medical products in Japan. These PAL revisions will regulate medical devices in Japan by promoting global harmonization and reinforcing safety measures. Among the changes are a new risk-based classification system, a Summary Technical Documentation (STED) format for the application process, a new ISO GMP regulation, and introduction of the market authorization holder (MAH).

Medical devices

The Japanese regulatory environment for medical devices has changed significantly since the enactment of the Pharmaceutical Affairs Law, which was implemented April 1, 2005, and further revised in 2006. The PMDA is an administrative agency that shares responsibility for medical device regulation.

Under the new PAL, the MAH system will greatly impact foreign medical device manufacturers entering the Japanese market or already established

Table 4.3 Organization of the Ministry of Health, Labor, and Welfare in Canada

Minister's Secretariat	Personnel Division, General Coordination Division, Finance Division Regional Bureau Administration Division, International Affairs Division, Health Sciences Division
Statistics and Information Department	Policy Planning Division, Vital and Health Statistics Division Social Statistics Division, Employment Statistics Division, Wages and Labor Welfare Statistics Division
Health Policy Bureau	General Affairs Division, Guidance of Medical Service Division, Medical Professions Division Dental Health Division, Nursing Division, Economic Affairs Division, Research and Development Division
Health Service Bureau	General Affairs Division, Specific Disease Control Division Tuberculosis and Infectious Diseases Control Division, Environmental Health Division, Water Supply Division
Pharmaceutical and Food Safety Bureau	General Affairs Division, Evaluation and Licensing Division Safety Division, Compliance and Narcotics Division, Blood and Blood Products Division
Department of Food Safety	Policy Planning and Communication Division, Standards and Evaluation Division, Inspection and Safety Division
Labor Standards Bureau	General Affairs Division, Inspection Division, Wages and Working Hours Division Labor Insurance Contribution Levy Division
Industrial Safety and Health Department	Policy Planning Division, Safety Division, Industrial Health Division Chemical Hazards Control Division Chemical Risk Assessment Office
Workers Compensation Department	Workers' Compensation Administration Division, Compensation Division Compensation Operation Office
Workers' Life Department	Policy Planning Division, Workers' Life Division General Affairs Division, Employment Policy Division
Employment Security Bureau	Employment Development Division, Employment Insurance Division, Public Employment Service Division, Private Employment Service Division, Foreign Workers' Affairs Division, Labor Market Center Operation Office

(Continued)

Table 4.3 (*Continued*)

Employment Measures for the Elderly and Persons with Disabilities Department	Policy Planning Division, Elderly Workers' Affairs Division Disabled Workers' Affairs Division
Human Resources Development Bureau	General Affairs Division, Human Resources Development Bureau Division Vocational Training Division, Vocational Ability Evaluation Division, Overseas Cooperation Division
Equal Employment, Children and Families Bureau	General Affairs Division, Equal Employment Policy Division Work and Family Harmonization Division, Part-time Work and Home Work Division, Family's Welfare Division, Child-rearing Promotion Division, Day Care Division, Maternal and Child Health Division
Social Welfare and War Victims' Relief Bureau	General Affairs Division, Public Assistance Division, Community Welfare and Services Division Welfare Promotion Division, Planning Division of War Victims' Relief, Relief Division, Record Division
Department of Health and Welfare for Persons with Disabilities	Policy Planning Division, Welfare Division for Persons with Disabilities Mental Health and Welfare Division
Health and Welfare Bureau for the Elderly	General Affairs Division, Long-term Care Insurance Division, Health Planning Division Promotion Division, Division of the Health for the Elderly
Health Insurance Bureau	General Affairs Division, Employees' Health Insurance Division National Health Insurance Division, Medical Economics Division, Actuarial Research Division
Pension Bureau	General Affairs Division, Pension Division Corporate Pension and National Pension Fund Division, Pension Fund Management Division, Investment Guidance Division, Actuarial Affairs Division
Director General for Policy Planning and Evaluation	Counsellor, Counsellor for Policy Evaluation

there. As of April 2005, the MAH system replaced the ICC (International Color Consortium) system. The ICC system required a manufacturer to obtain a license (*kyoka*) and approval (*shonin*) for each product. The manufacturer produced the devices and also released them onto the market. The MAH system will separate these two responsibilities: the production will be done by the manufacturer, and the MAH will act as an enhanced regulatory control mechanism to give final permission for product release. MHLW separated these two responsibilities to better regulate the quality and safety of devices.

All companies selling products in Japan are required to have a MAH. It must be located in Japan and be able to purchase or import products from a manufacturer, sell products to sales organizations, and, sometimes, temporarily store products in a licensed establishment. Companies have the option of using a distributor or importer, a third party, or designating themselves as the MAH. In this last case, a company may designate a subsidiary, branch, or representative office in Japan, as long as the MAH meets the aforementioned requirements. A MAH must designate up to three people as controllers, who will be responsible for overseeing product manufacturing, distribution, and release. The first controller is a general manager to oversee the overall product marketing, quality, and safety. The second, the quality assurance (QA) controller, is in charge of good quality practice (GQP). His or her responsibilities include ensuring that the manufacturer abides by the appropriate shipping and receiving methods, notifying MHLW of any changes in manufacturing or in-process controls, developing release criteria for each product, and handling communications for any recalls. The third controller is the postmarketing safety controller, who is accountable for good vigilance practice (GVP). This controller monitors the safety of products released onto the market, providing reports to the appropriate health authorities on any adverse incidents, recalls, and so forth. For class I products, only one controller is required to perform all of the functions. A MAH for class II products requires two people to carry out the functions. Class III and IV products must have three people, one designated for each role.

The PMDA arose from consolidation of the Pharmaceuticals and Medical Device Evaluation Center, the Japan Association for the Advancement of Medical Equipment (JAAME), and the Organization for Pharmaceutical Safety and Research. Primary responsibility of the PMDA includes overseeing the healthcare product approval by creating a streamlined and efficient review process and accelerating the approval cycle for medical devices. The PMDA has joined the Global Harmonization Task Force (GHTF) to develop a medical device review structure that incorporates the GHTF's agenda and recognizes medical device submission by the International Organization of Standards (ISO) and International Electrotechnical Commission (IEC).

MHLW

Significant changes have also been made to MHLW's preapproval and post-marketing system procedure and programs. The MHLW's recategorization of devices is based on GHTF's risk-based classification system as follows:

Classification of devices

- Class I devices are designated as general devices and require self-certification and no approval.
- Class II devices are designated as controlled devices and require certification and MHLW approval with review by a third-party body and PMDA.
- Class III and IV are designated as highly controlled devices and require MHLW approval with PMDA as the approval body.

For combination products, taking the drug-eluting stent (DES), the regulatory pathway in Japan was defined as a class III medical device.

Pharmaceutical and Medical Safety Bureau

In pursuit of safety of drugs and medical care, the bureau implements measures for securing the efficacy and safety of drugs/quasi-drugs, cosmetics, and medical devices, safety measures for medical institutions, and measures against narcotics and stimulants, while handling blood business. Thus, in light of the above, the Pharmaceutical and Medical Safety Bureau strives to protect people's lives and health through comprehensive efforts, covering clinical experiments and technical examinations for registration and after-sales follow-ups, to secure the safety and efficacy of pharmaceuticals directly related with people's lives and health. In addition, efforts for securing international harmonization among regulations on pharmaceuticals are being promoted, aiming to provide excellent drugs promptly to nationals through, for example, the International Conference on Harmonization (ICH) of Technical Requirements for the Registration of Pharmaceuticals for Human Use among Japan, the United States, and the EU. Japan's Pharmaceutical and Medical Devices Agency also intends to accelerate the drug-approval process by 2009 by means of a fast-track process for new drugs with significant clinical benefits.

Japan is the most regulated market, and consequently, it is a market having the highest cost of healthcare. Compared to the United States, products are usually sold at a higher price. For example, a certified mechanical bileaflet heart valve sells in India for (in U.S. dollars) $1,000, Europe for $4,000, the United States for $7,000, and Japan for $12,000. These increased regulations very often increased the average time of submission to approval of a new drug to about 17 to 20 months in 2003, compared to the United States and Europe, where the average time of submission to approval averaged 12 months.

Within the Ministry of Health, Labor, and Welfare (MHLW), the National Institute of Health Sciences (NIHS), which was established in Tokyo in 1874 as the Tokyo Drug Control Laboratory (later renamed the Tokyo Institute of Hygienic Sciences), is now a major organization. It is the oldest national research institute in Japan and currently consists of twenty-one divisions, five of which belong to the Biological Safety Research Center (BSRC). The major responsibilities of the NIHS involve extensive testing and research to ensure the quality, efficacy, and safety of chemical substances (including pharmaceuticals and food) that are closely related to people's lives. These activities can be broken down into the following fields:

1. *Pharmaceuticals and medical devices*: The NIHS is engaged in testing, evaluation, and related research aimed at maintaining the quality, efficacy, and safety of pharmaceuticals derived from chemical synthesis or organisms, biotechnology-based pharmaceuticals, therapeutics for gene or cell therapy, crude drugs, and medical devices.
2. *Foods*: The NIHS engages in research to establish standard methods for analyzing pesticide residues, veterinary drugs, allergic substances in foods, and food additives. It also conducts testing, research, and surveys to ensure the chemical safety of newly developed foods, food additives, food utensils, packages, and other items, while also conducting studies and testing to prevent health hazards caused by food-poisoning bacteria, microbial toxins, and so on.
3. *Other chemical substances* to which people are exposed: The NIHS is engaged in testing and research of household products, drinking water, indoor air, and so on, from a hygiene-chemistry standpoint. It is also involved in simultaneous pass/fail tests of cosmetics, preparation of codes and standards, and testing and research to evaluate the quality and safety of cosmetics and quasi-drugs.
4. *Biological studies*: The NIHS conducts testing and research on chemical substances in pharmaceuticals, foods, food additives, substances found in daily living, and so on, using experimental animals, tissues, and cells. It is also engaged in research on the establishment of testing and evaluation methods.
5. *Safety information on drugs, foods, and chemicals*: The NIHS collects information pertaining to the safety of drugs, foods, and chemical substances both in Japan and abroad, maintains its own database, and supports testing and research while also cooperating with international organizations. Development and maintenance of network systems to support the research activities of the NIHS. The NIHS takes part in projects related to the International Conference on Harmonization of Technical Requirements for Registration of Pharmaceuticals for Human Use (ICH) in Japan, the United States, and Europe.

Division of drugs

The division consists of three sections engaged in research work related to chemical analysis and pharmaceutical (biological and physicochemical) evaluation of drugs and their preparations, as well as evaluation of specifications submitted for new drug applications. In order to ensure the quality, efficacy, and safety of synthetic drugs, research activities of the division are focused on the following areas: (1) relation of genetic polymorph with drug disposition, (2) biopharmaceutical evaluation of drugs including bioequivalence tests, (3) evaluation of the physical and chemical stability of pharmaceuticals, (4) physicochemical characterization of dosage forms, (5) quality assurance system and validation for good manufacturing practices, and (6) new analytical methods. The division has contributed to establish drug standards and general tests in the Japanese Pharmacopoeia and other official books. The education and training of GMP inspectors is also a mission.

Division of biological chemistry and biologicals

The division consists of three sections responsible for the quality control, inspection, and evaluation of a variety of medical agents, including hormones, enzymes, proteins, biologically active high-molecular-weight compounds, and biotechnologically derived medicines. Biochemical studies on these drugs are also the subject of research activity. In addition, experimental preparations of newly developed drugs and standardization of assay methods are made.

Division of medical devices

The division, which has five sections, covers a wide range of products and materials as follows: medical devices (dialyzers, intraocular lenses, catheters, etc.), tissue-engineered medical devices, household products (including clothes, shoes, gloves, etc.), and their materials (PVC, rubber, polyurethane, silicone, etc.). Material-tissue interaction is the major subject of the research of the division. Implant data system and analysis of retrieval is another subject.

China

Medical equipment and device makers from all over the world have been focusing on the Asian markets for medical devices for a long time now. As a result of China's economic growth at a rate of greater than 10% annually, dramatic salary increases in the bigger cities, and following China's entry into the World Trade Organization (WTO), China's healthcare regulator, the State Food and Drug Administration (SFDA), has been making greater efforts to create a better regulatory environment. China is the second biggest market in Asia after Japan, and it has been improving its regulatory system over the past decade. This can be seen in the way China has implemented new laws to regulate drug prices and currently requires all drug manufacturers to be GMP certified. Both the medical device and pharmaceutical industries are growing at approximately 8–11% annually and are valued at more than US$25 billion. In May 2006, the SFDA issued project plans to revise the industry standards for medical devices. When complete, these revisions will simplify the device registration process in China.

Agencies and regulations

China's regulatory infrastructure is based on a traditional, bureaucratic model. There are several agencies involved in the regulation of medical devices, as described below:

- Ministry of Health (MOH)
- State Food and Drug Administration (SFDA), which is the Chinese equivalent of the U.S. Food and Drug Administration
- General Administration of Quality Supervision, Inspection, and Quarantine (AQSIQ)
- China National Certification and Accreditation Administration (CNCA)

The State Food and Drug Administration (SFDA)

In 2006, SFDA made significant efforts to improve medical device supervision and increase the efficiency of medical device testing in China. Currently the national testing centers in China perform tests on multiple device types, not only to increase efficiency but also to increase competition. Additionally, numerous medical device supervision and testing centers have also been set up. An intensive training program has also been implemented by the SFDA for the medical device testing officials. The national testing centers are accredited by the China National Accreditation Committee for Laboratories and the General Administration of Quality Supervision, Inspection, and Quarantine. The centers are responsible not only for supervising product quality, testing, and evaluation, but also for studying and proposing new technical standards.

The SFDA is responsible for overall supervision of drugs, food, cosmetics, and medical devices. Key responsibilities of the SFDA include the following:

- Drafting and enforcing laws and regulations
- Supervising safety management
- Investigating safety accidents
- Coordinating testing and evaluation activities
- Product registration
- Controlling the quality of medical devices

The SFDA department of medical devices

This division regulates medical devices for the central government. Responsibilities include drafting laws and regulations on device administration, device standards, and GMP; formulation of product classification lists; registration of imported medical devices and clinical trial sites; issuance of product registration certificates and production licenses; review of medical device advertisements; and controlling the quality of devices. The SFDA Medical Device Evaluation Center is also responsible for the technical review of devices. Responsibilities include:

- Review of domestic devices
- Review of imported devices
- Quality system inspections
- Review of clinical trial protocols

A three-tiered risk-based classification system is used by the SFDA as described below:

- Class I medical devices: Safety and effectiveness can be ensured through routine administration.
- Class II medical devices: Further control is required to ensure safety and effectiveness.
- Class III medical devices: Implanted into the human body, or used for life support or sustenance, or pose potential risk to the human body and thus must be strictly controlled in respect to safety and effectiveness.

The SFDA recently drafted a regulation called Criteria for Medical Devices to strengthen medical device quality. The draft integrates the ISO 13485 standard and the U.S. FDA quality system inspection technique along with the requirements of China. It requires manufacturers to document all design and development procedures for production, as well as record each batch of medical devices.

SFDA departments

There are several departments under the SFDA, as follows.

Department of Drug Registration Responsibilities of this department include drafting and revising national drug standards; formulating and updating product lists of immediate packaging materials and containers to drugs, and the requirements and standards for their medical use; taking charge of registering new drugs, drugs with national standards, import drugs, immediate packaging materials, and containers to drugs; implementing protection systems for traditional Chinese medicinal preparations; giving guidance to work of drug testing institutions nationwide; drafting criteria of marketing authorization for health food; and taking charge of approval of health food.

There are several divisions and one office under the department:

1. Division of Traditional Chinese Medicine
2. Division of Pharmaceuticals
3. Division of Biological Products
4. Division of Health Food
5. Division of General Management
6. Office for Acceptance of Drug Application

Classification of drugs In China, drugs are divided into three categories: (1) chemical drug, (2) biological drug, or (3) traditional Chinese medicine. Under these three categories, there are various classes. New chemical drugs have six classes, biological drugs have fifteen classes, and traditional Chinese medicine has nine classes. These classes mostly demonstrate and distinguish how a drug and its preparation process have been marketed, especially in regards to China. For example, class 1 of chemical drugs refers to a new drug that has never been marketed in any country, while class 3 of chemical drugs refers to a new drug that has only been marketed outside of China.

Department of Medical Devices Responsibilities of this department include drafting relevant national standards; drawing up and revising professional standards and good manufacturing practices of medical devices and medical dressings, and supervising their implementation; formulating the list of classified medical devices in consultation with the health authority under the State Council; being in charge of registration and regulation of medical devices; monitoring adverse events of medical devices; certifying clinical study bases, testing institutions, and quality system auditing institutions of medical devices; and being responsible for management of advertisement approval for medical devices.

There are three divisions and one office under this department:

1. Division of Standards
2. Division of Product Registration
3. Division of Safety Supervision
4. Office for Acceptance of Medical Device Application

Medical devices, as defined by these regulations, refers to any instrument, apparatus, appliance, material, or other article, whether used alone or in combination, including the software necessary for its proper application. It does not achieve its principal action in or on the human body by means of pharmacology, immunology, or metabolism, but may be assisted in its function by such means, the use of which is to achieve the following intended objectives:

1. Diagnosis, prevention, monitoring, treatment, or alleviation of disease
2. Diagnosis, monitoring, treatment, alleviation, of or compensation for an injury or handicap conditions
3. Investigation, replacement, or modification for anatomy or a physiological process
4. Control of conception

India

Like China, India is also implementing changes to its regulatory environment for pharmaceuticals, medical devices, and other healthcare products to meet the needs of a rapidly expanding economy and a 12% growth in its medical device market. However, unlike China, India is far from having separate departments to handle medical devices.

The Central Drug Standard Control Organization (CDSCO) in the Ministry of Health is the authority regulating products both drug and device products in India. The CDSCO lays down rules and standards and approves import and manufacturing of drugs, diagnostics, devices, and cosmetics. Currently, CDSCO's functions are to establish the standards and regulations for drugs, blood and blood products, intravenous fluids, and vaccines. With added responsibility of regulating the medical devices industry, CDSCO will be the approving authority for import, manufacture, and sale of medical devices in India. The regulatory procedure will be clear only after the government finalizes the regulations and the CDSCO provides the import guidelines.

Both the manufacturer and the importer will have to register with CDSCO. The Indian importer will have to obtain a "no objection" certificate to import and sell in India. It is expected, however, that for products that are approved by the FDA or CE, the registration process to obtain an approval to sell will be trouble-free. To register a new medical device or non-FDA/CE-approved devices in India, an application will have to be submitted to the regulatory authority along with documents such as details of the regulatory status in other countries, restrictions of use in approved countries, a free sale certificate from the country of origin, and so forth.

The Ministry of Health and F.W. under Gazette notification S.O. 1468 (E), dated June 10, 2005, declared the following sterile devices to be considered drugs under Section 3(b)(iv) of the act:

1. Cardiac stents
2. Drug-eluting stents
3. Catheters
4. Intraocular lenses

5. IV cannulae
6. Bone cements
7. Heart valves
8. Scalp vein set
9. Orthopedic implants
10. Internal prosthetic replacements

The Drug Controller General of India (DCGI) formulated guidelines for the import and manufacture of medical devices to be effective from June 26, 2006, with additional revisions. As per the guidelines, the above ten categories of sterile devices declared medical "drugs" under the Drugs and Cosmetics Act would be imported under the procedure for registration and import license prescribed by the same—a move that hoped to bring additional discipline to the sector, even though the primary mode of action of a medical device is not metabolic, immunological, or pharmacological. A point to note is that the drug-eluting stents, a combination product, was classified under regulations for drugs.

In 2004, the Mashelkar Committee called for the creation of a specific medical devices division within the CDSCO to address the management, approval, certification, and quality assurance of all medical devices in India. These regulations work within a similar framework as medical drugs and aim to enhance the requirements for devices that were subject to few or no controls, reduce duplication of devices previously assessed by foreign regulatory bodies, and place increased emphasis on manufacturer quality, risk management systems, and postmarket surveillance.

The regulation of medical devices

Currently, nonsterile devices can be freely imported, sterilized in India at a cheaper cost, and sold without any appropriate certification. The first step toward adherence of new policies is the publication and dissemination of guidance documents and public awareness programs.

A simple registration process for certified devices, proportionate registration costs, and constant update of regulations when required will go a long way in ensuring that only safe and effective medical devices are sold in India. Industry experts have recommended that all devices that are USFDA or CE certified should be fast-tracked and not have to go through the entire regulatory process in India, since the amount of regulations and cost of healthcare within a country are directly proportionate.

Information required to be included with the application for a registration certificate in India

1. Product information
    - Proprietary/brand name
    - Brief description of the device

- Category of the device
- Intended use and method of use
- Qualitative and quantitative particulars of the constituents
- Brief description of the method of manufacture and specification of the material used
- Contraindications, warnings, precautions, and potential AEs, wherever applicable
- List of accessories and other devices or equipment to be used in combination with the device
- Variations in shape, style, or size of the device, if applicable
- Labeling details conforming to the Drug and Cosmetics Rule (1945)
- Physician manual and promotional literature (literature insert) in English
- Packaging description, including pack sizes
- Recommended storage conditions
- Summary indications of any reported problems
- Details of standards to which the device conforms, along with a copy of the standards

2. Regulatory status
   - Approval of the product from any other regulatory agency (separate evidence of approval from each agency required)
   - U.S. FDA clearance/approval
   - EU Medical Devices Directive (CE certificate)
   - Australia/Canada/Japan approval
   - Approval in any other country
   - Copy of ISO/EN certification, if any, for the manufacturing facility
   - List of countries where the device is sold
   - List of countries where device is withdrawn from sale, with reasons, if any
3. Master file (details of GMP to ensure quality)
   - Component/material used
   - Device master file
   - Manufacturing process/flowchart
   - QA procedures/process controls
   - Final product testing or design inputs/outputs verification, if applicable
   - Functionality test protocol and report, if applicable
   - Risk assessment as per ISO 14971
   - Sterilization process and validation/verification
   - Stability data or statement of established stability of material used as applicable
   - Shelf life of the device
   - Biocompatibility and toxicological data, wherever applicable
   - Device GMP certificate

4. For combination products (e.g., devices containing medicinal product)
   - Data on safety, quality, and usefulness of the medicinal substance used
   - Data on compatibility with medicinal products, if device is intended to deliver medicinal products
   - Clinical data and published articles, if any
   - Batch release certificate for products incorporating any medicinal substance or substances of animal origin
   - For devices not approved for marketing in the country of origin, reports of clinical trials, details of sales, certificates of satisfactory use from medical specialists about the use of the device, and details of product complaints, if any

Medical devices with a prior approval from any of the recognized regulatory authorities, such as the United States, EU, Canada, Japan, and Australia, may be subjected to an abridged evaluation, and only a summary of all the studies and information described needs to be submitted in these cases.

For postmarket surveillance, the following information should also be included in the registration:

- Procedures for distribution of records
- Procedures for complaint handling
- Procedures for adverse incident reporting
- Procedures for product recall

In this chapter we tried to provide as much detail as possible to help companies understand the regulatory climate and guidance in various countries, including the United States, EU, Canada, Japan, and China, to gain approval and to market combination products. We urge the reader to pay close attention to evolving regulations in the area of combination products and various evolving requirements in other countries, as the export of these products from the United States and Europe will be on the rise in the future. One recommendation we can provide to companies focusing on combination products is to create an overall regulatory plan first. This plan shall focus on gaining approval for the combination products in the United States and Europe following FDA and EU regulations, respectively, for product approval prior to any other country. Plans to gain approval in other countries shall look for leveraging the FDA and EU approval efforts since the regulatory pathways for combination products in the United States and EU are fairly similar and well defined. In general, products that are similar to approved combination products will be easier to develop and launch into market since the challenges and knowledge can be leveraged off the already approved product and its regulatory history with the FDA.

Bibliography

21 CFR, Part 3. *Product jurisdiction*, November 21, 1991.

21 CFR, Part 210. *Drug cGMP regulation*, 2006.

21 CFR, Part 211. Drug cGMP regulation, 2006.

21 CFR, Part 314. *Applications for FDA approval to market of a new drug, 2006.*

21 CFR, Part 820. *Design controls for medical devices*, 2006.

21 CFR, Part 820. *Quality system regulations*, 2006.

Application of cGMP regulations to combination products: frequently asked questions. Guidance for industry and FDA staff, Office of Combination Products, January 15, 2007.

Application user fees for combination products; availability. Draft guidance for industry and FDA staff. *Federal Register* 69(87):57942–57943, 2004.

Assignment of agency component for review of premarket applications. Final rule, FDA. *Federal Register* 68:37075–37077, 2003.

China food and drug regulatory website: www.sfda.gov.cn

Combination products: timeliness of premarket reviews: dispute resolution guidance. Draft guidance for industry, FDA, 2004. www.fda.gov/oc/combination/dispute.html.

Cramer, Chris and Rastogi, Sharad. Drug and device combinations, combination medical products: capitalizing on convergence. *Medical Device & Diagnostic Industry*, January 2007.

Current good manufacturing practice for combination products. Draft guidance for industry, FDA, 2004. www.fda.gov/oc/combination/OCLove1dft.html.

Definition of primary mode of action of a combination product. Proposed rule, FDA. *Federal Register* 69(89):25527–25533, 2004.

DiMasi, Joseph, et al. The price of innovation: new estimates of drug development costs. *Journal of Health Economics* 22: 151–185, 2003.

Early collaboration meetings under the FDA Modernization Act: http://www.fda.gov/cdrh/ode/guidance/310.html and final guidance for industry and CDRH staff: http://www.fda.gov/cdrh/devadvice/ide/print/approval.html

Early development considerations for innovative combination products. Guidance for industry and FDA staff, Office of Combination Products, September 2006.

Innovative systems for delivery of drugs and biologics: scientific, clinical and regulatory challenges. FDA workshop. Bethesda, Maryland, July 8, 2003.

Federal *Food, Drug, and Cosmetic Act*, Section 503(g), as amended: http://www.fda.gov/opacom/laws/fdcact/fdctoc.htm

Federal *Food, Drug, and Cosmetic Act*, Section 563, as amended: http://www.fda.gov/opacom/laws/fdcact/fdctoc.htm

Final rule on the definition of primary mode of action of a combination product. *Federal Register* 49848, August 25, 2005.

Fox, David M., and Shapiro, Jeffrey K. Combination products: how to develop the optimal strategic path for approval. *FDA News*, 2005.

Gross, Ames. Growing opportunities in the Indian medical device market. *Medical Device & Diagnostic Industry*, May 1995.

Gross, Ames and Loh, Nancy. Medical device regulatory update: China and Japan. *Medical Device & Diagnostic Industry*, October 2006.

Gross, Ames and Weintraub, Rachel. Regulatory updates for drugs, devices & IVDs in Asia. *Regulatory Affairs Focus*, May 2005.

Guidance for Industry, Investigators, and Reviewers Exploratory IND Studies. CDER (Pharmacology/Toxicology), January 2006.

http://www.competitionbureau.gc.ca/internet/index.cfm

http://www.fda.gov/oc/cdrh/devadvice

http://www.fda.gov/oc/ombudsman/drug_dev.htm

http://www.hc-sc.gc.ca/ahc-asc/branch-dirgen

http://www.hc-sc.gc.ca/dhp-mps/prodpharma/applic-demande/pol/combo_mixte_pol_2006_e.html (last accessed on August 15, 2006)

Indian drug regulations: http://www.cdsco.nic.in/

Inose, Christine and Brown, Meredith S. Combination products in the US: navigating the regulatory jungle. *Regulatory Affairs Focus*, pp.17–22, May 2002.

Intercenter Agreement between the Center for Biologics Evaluation and Research and the Center for Devices and Radiological Health, October 31, 1991: http://www.fda.gov/oc/ombudsman/bio_dev.htm

Intercenter Agreement between the Center for Drug Evaluation and Research and the Center for Devices and Radiological Health, October 31, 1991: http://ec.europa.eu/enterprise/medical_devices/meddev/2_1307-2001.pdf (accessed June 30, 2006)

Jepson, Gordon. The business impact of noncompliance (consequences of noncompliance in Canada: regulatory enforcement agencies, powers and practices). *Regulatory Affairs Focus*, January 2007.

Kader, Victoria and Priestley, Duaine A. India's medical device market is becoming too big to ignore. *Medical Device & Diagnostic Industry*, April 1997.

Leichter, Lee H. Pathways and strategies for combination products in the EU. *Regulatory Affairs Focus*, September 2003.

Office of Combination Products, FDA homepage: http://www.fda.gov/oc/combination/

Outcomes of the summit on regulatory issues involving combination products. White paper prepared by RAPS, July 2005.

Patel, Neesha. Implications of medical device regulations. *Express Health Care Management*, September 2006.

Portnoy, Stuart and Koepke, Steven. Regulatory strategy: drug testing of combination products. *Medical Device & Diagnostic Industry*, June 2005.

Stephens, Robin N. Bringing drug-eluted stents to Europe. *Regulatory Affairs Focus*, pp. 57–59, February/March 2004.

chapter five

Resource requirements

In Chapters 3 and 4, we presented not only the need for a rigorous approach in developing combination products and gaining regulatory approval for them, but also the knowledge necessary to accomplish those. Since the development of combination products must critically consider other requirements, this chapter takes into consideration not only the technological challenges but also a much-needed focus on the development and implementation of support systems for combination products (CPs). These include, but are not limited to, IT resources, technical resources, facilities, quality, regulatory, clinical, and manufacturing system challenges.

As we mentioned in earlier chapters, drug-device and other combination products are a rapidly growing area in today's environment within the healthcare technology. However, these new technologies for combination products pose many challenges for the traditional research, manufacturing, quality, and regulatory departments as they strive to understand and address the requirements of two separated components of a combination product that are Food and Drug Administration (FDA) regulated. It is not uncommon to hear the following questions:

Who is responsible for clinical trials? Is it us or the company that developed the drug for this product?

Do we need design control systems in addition to our current cGMP system?

Will there be a preapproval inspection for this product?

Developing combination products poses a range of unique hurdles covering many areas, such as business challenges and technical, regulatory, and quality uncertainties due to the uniqueness of components coming together from a biologic, pharmaceutical, or device family to form the combined product. It is mandatory and key to have input from all perspective partners in the early phases of project planning, as a developer with strong previous experience in the device area may have a tendency to minimize drug or biologic issues; similarly, a developer with expertise in the drug areas may likely overlook the device requirements. Refer to chapter VII, which also covers some of the issues in the development and manufacture of combination products.

Resources

A significant organizational challenge faced by companies in the development of combination products is acquiring or integrating individuals with specific skill sets to enhance the employee resource pool. These companies need to initially identify the correct skill sets, current capabilities within the organization of individuals, resources that can be leveraged from collaborative partners, and so on. To ensure a successful team structure for combination product development, the organization needs to address specific requirements that may be lacking within an organization. Hence, it is imperative for the organization to build the correct skill sets and capabilities within the organization early on in development.

First and foremost, a capability that needs to be built is having the right leadership. This includes general management as well as product development leadership. Since combination product development, manufacturing, marketing, sales, and service are very involved and can be frustrating at times, the company leadership should have a lot of patience and be committed. They should be willing to learn topics that are outside their core expertise. For example, in the case of a drug-coated stent, the leadership of the company should be willing to learn topics related to drug development and marketing. This will prevent unrealistic expectations on time to market, time to break even, revenue forecast, and so on. They should also have the ability to foster collaboration at all levels.

In addition to having the right leadership, combination product work requires the right competency at all levels. This starts early during the development of the product. An initial task for a company during the early product development stages is to document a developmental strategy for the novel combination product. As a result, significant gaps in the organizational skills and capabilities would be identified.

An example can be seen with a company X, which is a device manufacturer with little to no biologic or pharmaceutical experience in development challenges of combination products that may be drug-device or biologic-device. In this case, company X needs to collaborate with partners to obtain any required skills that are missing within its organization. Failing to do so may result in many undesirable consequences, including, but not limited to:

- Development of combination products may be delayed or cancelled due to high initial investments.
- Incorrect drug-biologic selection and proof concept for a device combination product.
- Inadequate or absence of unique technical skills in organizations developing combination products.

Table 5.1 lists some of the expertise that is typically required for combination product development.

In order for a company to successfully develop and launch combination products, the company may need to restructure its organization in skill sets, leaders, and core strategies if developing combination products is going to become a mainstream business for it. To ensure success, such organizations need to align their organizational structures and product development

Table 5.1 Examples of Expertise Requirements Unique to Combination Products

Product and processes development
Technology transfer and scale-up
Pricing and reimbursement strategy
Manufacturing and supply chain strategy
Analytical methods, quality systems, and regulatory strategies
Preclinical and clinical expertise
Specialized marketing expertise
Organizational development
Project management

processes and systems, incorporating an integrated approach of a device-pharmaceutical model for a device-drug combination product.

Addition of a drug or biologic to a device

Device manufacturers must carefully evaluate their in-house research and development (R&D) capabilities. It is critical to broaden technical skills to understand the implications of biopharmaceutical development issues, such as how the design affects the ability to coat the device, or how changes to the product or manufacturing process influence stability. As a result, device manufacturers often add new skills and capabilities that did not exist in the device-only world. Additionally, they must strengthen the technical skills of their quality, regulatory, and clinical organizations to deal with pharmaceutical or biologic issues. Identification of the proper skill sets and capabilities must be addressed throughout the organization, including management, project teams, and across functions.

Device companies need to take into account what the user needs, and intended uses are from the drug component of the combination product. The requirements have to be translated into new technical design inputs that define the drug development areas, such as pH, dosing concentrations, and elution profiles. Often, with these types of combination products, the internal manufacturing capabilities and constraints may significantly impact in-process, final product, and shelf life of the combined product.

For the combination product, the potential interaction between the pharmaceutical/biological compound and the device constituent needs to be assessed in the early phase of drug development. The following areas need to be looked at:

1. Technology and capabilities assessment needs to be conducted to ensure understanding of the pharmaceutical/biological constituent.
2. The impact of the technology/processing on the active product intermediate (API), the chemical entity that has drug activity and structure

but is not yet formulated with excipients, has to be evaluated in understanding the application of the drug/biologic to the device.
3. Analytical and assay test methods have to be understood, and the stability testing and strategy need to be considered.
4. The toxicology, pharmacology, and safety performance of the drug/biologic needs to be understood.
5. Timing issues need to be assessed when looking at new components versus line extension.
6. Facilities, utilities, and environmental needs have to be assessed for the potential drug/biologic.
7. Skill sets of individuals that need to address the regulatory, analytical, quality, and clinical needs of the project have to be assessed. Additionally, in order to successfully develop a marketable product, it is extremely important that device companies engage the pharmaceutical subject matter experts (i.e., regulatory, clinical, quality operations, and marketing) in product development throughout the development cycle.
8. Additional discovery work on the drug/biological product may be required if the drug exposure to the device and the dose range are different from those previously studied.
9. Biologics are very labile substances, and their viability, stability, and effectiveness can easily be altered by temperature, agitation, and materials with which they come in contact; these parameters have to be reviewed by the team.

Additionally, device companies need to also consider the following key good manufacturing practices (GMP) provisions during and after joining together copackaged and single-entity combination products:

- Calculation of yields
- Testing and approval or rejection of components
- Drug product containers and closures
- Expiration dating
- Testing and release for distribution
- Stability testing
- Special testing requirements
- Reserve samples

Addition of a device to a biologic-device combination product

In this case, companies need to have regulatory experts from both drug and device sectors and need to engage in early and frequent interactions with both FDA centers. Establishing a relationship with the Office of Combination Products (OCP) can be critical if questions/issues arise that cannot be resolved with the FDA lead centers (CDER, CBER, CDRH, etc.). The OCP should be used to set up meetings and be the liaison between industry and the lead center.

Consultation or collaboration between centers on product reviews has been traditionally performed. Every effort should be made to identify the need for a consultative or collaborative review as early in the review process as possible. Frequent dialogues with the identified review centers are also necessary in order to address issues.

The submission needs to incorporate the requirements for both investigational new drug (IND) and investigational device exemption (IDE) requirements, as in Table 5.2.

Table 5.2 FDA Requirement versus Documentation in the Submission

FDA requirement	Documentation to be included in submission
IND/IDE	1. An approved BLA or premarket approval application (PMA) for other uses can be referenced. 2. One also needs to confirm with both centers of the FDA to ensure their requirements are met.
Clinical evaluations	Evaluations of investigational devices, unless exempt, must have an approved IDE prior to study.
FDA response	Following the IDE route, an FDA response must be received in writing before proceeding. The response can be conditional approval, approved, or not approved.
Based on risk assessment of a device	There are different types of IDEs: 1. A traditional IDE for a significant risk device 2. An abbreviated IDE for a nonsignificant risk device 3. A treatment IDE for seriously ill/life-threatening disease
GMP provisions	Biologics companies need to consider: 1. Development of design control procedures and institute design controls 2. Inclusion of stage gate and design reviews at designated points in development
Purchasing controls	1. Develop procedures for purchasing controls to govern material suppliers, contractors, and consultants. 2. Purchasing control should begin once a project team leaves the feasibility stage. Timing this with the start of design control is advised.
Corrective/preventive action program (CAPA)	1. A CAPA must be implemented. 2. The CAPA system has to include a method for determining the effectiveness of corrective actions implemented. Both preventive and corrective actions must be documented and their effectiveness measured.
Management review	The management review process is intended to assess the effectiveness of the quality system for the entire company. Management review should be conducted at least annually to assess the health and effectiveness of the company's quality system.

Additionally, if a device is added to the biologic product in developing a combination product, the team also needs to ensure effective supplier and external manufacturer selection, develop an effective and efficient supply chain, and ensure all processes are validated.

Facilities

In a drug-device combination product, the manufacturing operations of a device company may very likely need to be updated to comply with the pharmaceutical current good manufacturing practice (cGMP) requirements for manufacturing the drug component. Additionally, partner companies can leverage off the manufacturing facilities at respective sites to launch the combination product. Manufacturing considerations such as scale-up and facility management in terms of compliance and quality are critical in the development of a combination product. During the development or enhancement of facilities for combination products, an organization needs to consider factors such as the stability of a combination product as a whole and the separate constituent parts during manufacture of the finished combination product. The facility has to be environmentally controlled to store and develop such products in terms of temperature, humidity, pest control, and environmental monitoring to ensure clean-room requirements are met during the manufacturing cycle of the combination product. For manufacture of constituent parts that utilize aseptic manufacturing processes, the facility is adequately designed to perform these various steps. Additionally, the organization needs to ensure that any drug or biological product constituent parts used in the combination product are not destroyed or changed during terminal sterilization techniques.

In-house testing

Medical device firms do not usually perform extensive chemical analysis for incoming components. The drug GMP requires identity testing plus conformity testing for purity, strength, and quality parameters. These tests can be performed in-house or by a contract lab. In lieu of conformity testing for these parameters, the manufacturer may accept a certificate of analysis from the supplier of the component. However, the certificate's reliability must be verified and monitored at appropriate intervals. To do this, the in-house or contract laboratory must confirm the testing results in a predetermined number of components and then repeat this step at a later date. A program to confirm the credibility of the certificate of analysis will need to be considered in the project timeline for the combination product if the medical device firm does not test for the purity, strength, and quality parameters of the incoming components.

Developing analytical laboratory capabilities

The medical device company needs to put in place an analytical laboratory, which is a significantly expansive operation in terms of purchase of specialized analytical equipment, finding contract partners or setting up new custom-built facilities, and recruiting technically competent analysts.

The analytical laboratory has significant responsibilities, which begin at product development, and is continuously needed through routine product-related testing and failure investigation. Test methods for components, in-process materials, finished product, cleaning residuals, and stability during the product development phase are validated by the analytical lab during R&D. The quality assurance function of the analytical laboratory tests these same materials during the development and commercialization phases of the product life cycle. During product failure investigation, the laboratory can also help to determine whether the problem was a result of processing, material, or laboratory error. Controlling materials, ensuring timely results, and preventing mix-ups during these three activities require experienced staff with expertise in documentation and material control skills during the processing of the combination product.

Clean rooms

The following areas need to be addressed to set up clean rooms in the manufacture of a biologic-pharma-device combination product. Both the design and qualification of the equipment and facility need to take place. Equipment has to be procured and installed, and the facility has to be designed to meet the clean-room requirements based on the classification of the clean room.

The development and scale-up of cell culture process or development and the scale-up of a purification process need to occur prior to transfer to manufacturing.

Manufacturing operations

In manufacturing of product for use in humans, there are regulatory challenges and concerns common to all biologics and drugs, such as product safety and characterization, control of the manufacturing process, reproducibility, and production lots being consistent. By following the QS and cGMP regulations applicable to drugs and devices, ones can ensure control of both the final product and the manufacturing process for the medical device company.

To comply with the pharmaceutical GMP requirements, a medical device manufacturer will very likely need to modify some of its existing operations to meet pharmaceutical GMP regulatory requirements. Defined and designated areas for quarantine goods, such as incoming components or in-process materials awaiting disposition by quality control, may be needed in manufacturing operations. Similarly, other designated areas to store rejected components or finished goods until they are either returned to the vendor or destroyed, as appropriate, may also be required.

Other factors that need to be considered include the facility water, which must be tested according to Environmental Protection Agency Primary Drinking Water Standards. Water may be used for washing containers, operator hands, or equipment. Therefore, water from various locations will need to be routinely sampled and tested.

Equipment and utensils used in drug manufacturing must not contaminate the drug coating or polymer solution. Contamination may occur when equipment and utensil surfaces, such as surfaces made of oxidative metals, react with the drug coating/polymer solution, or when other surfaces, like plastics, absorb the drug. Some solvents used in drug formulation can react with rubber, so sampling devices like syringes with rubber plungers should be avoided. Manufacturing, quality, and maintenance staff should know which materials should not come in contact with the drug coating or polymer solution.

Comprehensive cleaning procedures need to be defined and validated. When not properly cleaned, equipment and utensils can become contaminated with carryover material from previous batches of the drug coating or polymer solution. Cleaning test methods must be validated to show they effectively remove contamination to below the maximum acceptable level established by the firm. To minimize equipment cleaning in production, some pharmaceutical firms allow less extensive cleaning between batches of identical product during manufacturing campaigns because they have shown through validation studies that residual carryover is below acceptable levels.

Most medical device manufacturers do not give due consideration to time limits for specific production processes. However, because of potential chemical instability, solvent evaporation, or microbial degradation of in-process materials, some processes may indeed require defined time limits. Unfortunately, the need for time limits, ideally set during the R&D phase, may not be discovered until a product fails during the commercialization phase.

The pharmaceutical GMP regulation has checks and balances that are meant to preclude or correct improper processing. For example, the regulation requires the manufacturer to determine actual yields and percentages of theoretical yields at the conclusion of appropriate phases, because actual yields provide an early warning of process problems. By including these yield checks throughout the process, the manufacturer can investigate problems in a timely manner, rather than waiting until the batch has been completed. A second person must verify the accuracy of yield calculations. Yield requirements are not mentioned in the device quality system regulation (QSR).

Since there are no specific models of the development processes for combination products yet, companies identify a model that best fits their specific combination product and customize the model to their specific needs and expertise.

An example in how companies may use different ways to formulate their development road map can be seen as follows: Company A may use the QSR model to describe its key requirements for device development initially, and subsequently integrate the relevant biologic or pharmaceutical requirements. The QSR describes general processes for the design, development, and testing of medical device products. For the pharmaceutical element of the combination product, cGMP are used, and the final output from the development team for transfer of the product from development to manufacturing is an integrated approach using elements from both sets of FDA regulations (the QSR) and cGMP requirements. This approach to integrated processes during development can successfully create stable processes and road maps for specific companies' development and integration of quality systems for combination products. Medical device companies can use this framework to design and implement a single, comprehensive system for combination product development. Such a framework can integrate relevant pharmaceutical practices with its design control requirements.

Quality

A GMP-compliant quality system for a combination product

For companies that are developing these combination products, taking the drug-coated cardiovascular stent as an example, which is now a product comprised of a device and a drug component, one of the key challenges is how to transform the company's current quality system into an effective quality system that would meet and comply with the FDA requirements for the novel combination product.

The FDA's Office of Combination Products (OCP) in 2002 provided a guidance for cardiovascular drug-eluting stents that suggested that "even though the agency has determined that these products would be subject to premarket review and approval solely under the medical device provisions of the Act, the agency has applied human drug current Good Manufacturing Practices (GMP) to the manufacture of the drug component of the combination product, and additionally may apply other drug requirements to the products as appropriate." As a result, manufacturing operations associated with the drug coating needed to comply with the pharmaceutical GMP regulations, 21 CFR Parts 210 and 211. The manufacture of the cardiovascular stent and its associated delivery system needed to comply with the QSR, 21 CFR 820 (device GMP). The challenges for these companies are evident in how the requirements for both the medical device and pharmaceutical regulations could be incorporated into one quality system for manufacturing the finished combination product (cardiovascular stents).

Cost and financial factors

In order to improve the cost and yield of combination products, organizations need to improve the determination of design requirements in the early phases. There is also a strong need for these organizations to not only develop, but also validate, the critical processes and methods early on. The transfer of technology to operations also needs to be improved and defined in the early phases of development.

Development of combination products is usually more costly than the development of medical devices, pharmaceuticals, or biologic products alone due to several unknowns at the onset of the project. These unknown factors include accurate prediction of the actual cost and financial resource requirements for the combination product.

The true cost is difficult to predict due to the unknown timeframes for delivery of such products from concept to commercialization. Companies need to evaluate realistic timelines, skill set requirements, developmental requirements, and compliance, quality, regulatory, and manufacturing requirements. In many instances, if the initial timeframes and resources are not managed well, the company's attempts at developing and subsequent launch of the product may fail. The device or pharmaceutical company looking to develop a novel combination product has to evaluate the organization's capability and product development processes to see if they are capable of meeting the challenges of the combination product technology from a design and development perspective. The organization has to align the developmental processes from both components of the combination product; for example, if the product is a device-drug combination, an integrated developmental process with the specific requirements of quality from both QSR and cGMP is developed.

Additionally, the developmental team and the supporting partners from regulatory, quality, clinical, and manufacturing need to be aligned in the requirements and expectations for the combined product and trained on both sets of requirements.

Companies should also explore the possibility of achieving tax benefits by developing these products in a tax-favorable nation. It is our recommendation that companies approach this area with extreme caution since tax advantage does not always mean the availability of competent and skilled workforce.

Fundamental differences in regulations between pharmaceuticals and medical devices

There are some major differences in the way the drugs and medical devices are regulated for an effective quality system to be developed and implemented for manufacture of these products, due to differences in the physical properties for drug components versus medical device components.

Drug components such as reagents or solvents are usually less stable than medical device components. Physical changes due to heat, light, humidity, microbial action, chemical action, or contamination may expose the drug components to degradation.

In contrast, medical device components are in general inert for egs, as they consist of electronic and metal parts. The drug GMP regulation therefore has more detailed and stringent requirements for control and monitor of environmental conditions, microbial contamination, chemical contamination, and material handling than its medical device QSR.

Expiration dating is a significant factor in control of cGMP for pharma in order to ensure the drug quality; labile drug components, in-process materials, and finished products usually require expiration dating. The initial drug component's expiration date is provided by the manufacturers of the drug (chemicals); however, expiration dates for in-process materials and finished products for the combination product are provided by the combination product manufacturer by means of conducting stability testing using analytical testing in a laboratory. Stability testing usually needs to commence at research and usually continues through the commercialization of the product. Stability testing and the analytical portion of the tests are time intensive and would likely affect the timeframes for launch of the product.

The medical device manufacturer does not have to perform such stability testing on devices; hence, stability testing for components, in-process materials, and finished products for the drug portion of the combination product need to be factored into the development project plan of the combination product.

The appearance of many drug components is similar, that is, they are white powders or clear liquids, so mix-ups during formulations could happen easily due to the similarity in appearance of these components. Additional verification of critical formulation steps could reduce the risk of this mix-up. Duplication checks in the device world are not usually the norm, as a sampling plan usually takes care of assessing the quality of the product during its manufacture.

The stability program

For the combination product, the stability program would require significant resources. Initially, a knowledgeable staff needs to be resourceful and dedicated to define the stability program. A reliable stability-indicating assay that can differentiate between the active ingredient and its impurities needs to be developed once the degradation of the drug component is known. Other stability assays that indicate identity and quality of the product may also be required. Competent resources in terms of staff are also required for acquiring samples, storage of these in stability chambers, and removal at the appropriate time for testing. These stability expenses are new for a medical device firm and must be factored into the total project cost.

Additionally, an analytical test method must be developed that is meaningful and representative of the drug's elution properties in the patient. The test method should be easily usable in a manufacturing setting in that the analysis must have a reasonable completion time and be robust enough for multiple analysts to reliably perform.

Regulatory approval process

In general, device companies prefer to go the route of a device approval process as first choice when presenting a combination product to the FDA. There are several advantages in this route, as the approval route usually includes less intensive preclinical testing, a single pivotal human clinical trial rather than multiple trials, and the application filing fees are considerably lower. However, there are other factors that also need to be considered, such as additional preclinical testing based on product risks, which can always be requested, even if the CDRH has primary jurisdiction for a particular combination product. Product evaluation and required testing may be necessary to fully understand the risks. The evaluation needs to consider both device-drug and device-biologic components based on the combination product.

Manufacturing regulations are also a consideration since, if a particular combination product is developed in a device environment with all the QSR procedures in place and implemented, the combination product should be easily integrated into the system with additional quality requirements from the cGMP. However, if the combination product is developed and produced in a pharmaceutical environment, the QSR requirements in particular reference to compliance with design controls have to be considered. Addition of an integrated system, training staff, and so on, can be costly and regulatory requirements need to be timely.

Bibliography

21 CFR, Part 210. *Drug cGMP regulation*, 2006.

21 CFR, Part 211. *Drug cGMP regulation*, 2006.

21 CFR, Part 820. *Quality system regulations*, 2006.

Application user fees for combination products. Guidance for industry and FDA staff. Draft guidance, FDA, September 2004.

Browne, Justin A. Ensuring accessibility of combination products. *Regulatory Affairs Focus*, January 2004.

CBER guidance webpage: http://www.fda.gov/cber/guidelines.htm

CDRH guidance webpage: http://www.fda.gov/cdrh/guidance.html

Current good manufacturing practice for combination products. Draft guidance for industry, FDA, 2004. www.fda.gov/oc/combination/OCLove1dft.html

Current good manufacturing practice regulations for combination products. *Federal Register* 22565, April 24, 2006.

Device advice webpage: http://www.fda.gov/cdrh/devadvice/

Early development considerations for innovative combination products. Guidance for industry and FDA staff, Office of Combination Products, September 2006.

Final rule on the definition of primary mode of action of a combination product. *Federal Register* 49848, 25 August 25, 2005.

Gibbs, Jeffrey, Phelps, Hyman and McNamara, P.C. State of the union: drug-device combinations. *Medical Device & Diagnostic Industry*, November 2006.
Kramer, Mark D. Combination products: challenges and progress. *Regulatory Affairs Focus*, pp. 30–35, August, 2005.
Mannix, Daniel G. Combination products. *Regulatory Affairs Focus*, September 2006.
Michalik, Robert J. Regulation of combination products: marmonization or unification? *Regulatory Affairs Focus*, September 2006.
Office of Combination Products, FDA homepage: http://www.fda.gov/oc/combination/
Proposed rule 887: current good manufacturing practice for combination products. *Federal Register* 22565, April 24, 2006.
Swain, Erik. Extra effort: packaging for combination products. *Medical Device & Diagnostic Industry* 27, pp. 98–106, 2005.

chapter six

Manufacturing of combination products

The Food and Drug Administration's (FDA) biologics regulations define *manufacturer* as "any legal person or entity engaged in the manufacture of a product subject to license under the PHS Act," including "any legal person or entity who is an applicant for a license where the applicant assumes responsibility for compliance with the applicable product and establishment standards" (21 Code of Federal Regulations (CFR) 600.3(t)). A manufacturer thus includes a license applicant, who may or may not own the facilities engaged in significant manufacturing steps, when such an applicant assumes responsibility for compliance with the applicable product and establishment standards, including, but not limited to, 21 CFR Parts 210, 211, 600 through 680, and 820. *Manufacture* is defined as "all steps in propagation or manufacture and preparation of products and includes but is not limited to filling, testing, labeling, packaging, and storage by the manufacturer" (21 CFR 600.3(u)). A manufacturer of a biological product must demonstrate responsibility for the manufacturing process as described in its biologic licence application (BLA) (21 CFR 600.3(t)). For example, a manufacturer must avoid introduction of contaminants during production (21 CFR 610.13).

It is well known that the FDA has established regulations, similar to the biologics regulations, for medical devices and pharmaceuticals. So it is natural to expect regulations to guide the manufacture of combination products. The FDA's current thinking about good manufacturing practices for combination products is described in *Draft Guidance for Industry and FDA Staff: Current Good Manufacturing Practice for Combination Products*, available at http://www.fda.gov/oc/combination/OCLove1dft.html.

The FDA has not promulgated current good manufacturing practice (cGMP) regulations specifically for combination products. Until it does so, each constituent part (i.e., the drug, device, or biological product) remains subject only to its governing current good manufacturing practice regulations when marketed separately (see 21 CFR 3.2(e)(3) and (4)) and when manufactured separately as constituent parts of a combination that will later be combined (see 21 CFR 3.2(e)(1) and (2)). For example, if a drug is marketed that is intended for use only with an approved individually specified device that is also marketed separately, the drug constituent must comply only with 21 CFR Parts 210 and 211, and the device constituent must comply only with 21 CFR Part 820. Similarly, during the time of separate manufacture (i.e., before drug and device combination products are produced as a single entity or are copackaged), 21 CFR Parts 210 and 211 apply only to the drug constituent, and 21 CFR Part 820 applies only to the device constituent.

However, for combination products that are produced as a single entity or are copackaged (see 21 CFR 3.2(e)(1) and (2)), both sets of current good manufacturing practice regulations are applicable during and after joining the constituent parts together.

The FDA recognizes that many manufacturing facilities operate under one type of current good manufacturing practice system (i.e., either that described by the quality system (QS) or that described by the cGMP regulation). As noted above, the FDA recognizes that there is considerable overlap between the QS and cGMP regulations. It should generally not be necessary for manufacturers who make combination products that are produced as a single entity or are copackaged to maintain two separate manufacturing systems to ensure compliance with both sets of regulations during and after joining the constituents together. The FDA believes that compliance with both sets of regulations during and after joining these types of combination products can generally be achieved by using either the cGMP or QS regulations, for example, by using the current good manufacturing practice system already operating at a manufacturing facility.

Manufacturing considerations

Manufacturing, scale-up, and quality management are important considerations during the development of a combination product. Manufacturing methodologies affect both premarket development and postmarket regulation. The FDA encourages consideration of the manufacturing issues posed by the scientific and technical aspects of the drug, biological product, and device constituent parts, and of the combination product as a whole. The FDA also encourages developers to carefully consider the effect of the manufacturing methods on the interaction of the constituent parts. For example, the stability of a combination product as a whole may be different than that of the separate constituent parts. Certain drug or biological product constituent parts may be altered or destroyed by terminal sterilization techniques. For constituent parts that use aseptic manufacturing techniques, developers are encouraged to implement manufacturing techniques to ensure aseptic control for the combination product.

During premarket investigation, once the preclinical and clinical studies begin, any potential change in the manufacturing process for the drug, biologic, or device constituents or for the combination product may affect the safety or effectiveness of the combination product as a whole. For example, changes in concentration, inactive ingredients, software, or the methods to combine two constituent parts could affect the performance characteristics of the combination product. When applying cellular constituent parts to a device, the performance characteristics may vary with the time and methods used for cellular incubation before application to the device constituent. Additionally, the applicable device constituent design controls would consider anticipated manufacturing changes during investigational development. In order to address such manufacturing considerations, it may be necessary to develop new manufacturing techniques, in-process testing, testing specifications, and other characterization methods to assess changes in the constituent parts and for the combination product as a whole.

For certain developmental changes, additional bridging studies (*in vitro*, preclinical, or clinical) may be appropriate. In addition to considering manufacturing changes that may occur during premarket development, the FDA also recommends early consideration of anticipated postmarket manufacturing changes for the combination product or its constituent parts. The FDA encourages manufacturers to establish arrangements with the manufacturers of constituent parts to maintain sufficient awareness of manufacturing changes in constituent parts that may occur during the premarket or postmarket period. Such awareness could help to ensure continued safety and effectiveness of the combination product by ensuring that the potential impact of a manufacturing change is evaluated in a manner appropriate for the stage of combination product development. As appropriate, these postmarket manufacturing changes may require careful review, validation, and prior approval before marketing. For some products, it may be helpful to develop postapproval change protocols for further discussion with the agency.

Investigational or marketing applications often contain trade secret or confidential commercial information. In some instances, developers may wish to provide all necessary information in one marketing application. However, for combination products developed by more than one manufacturer, there may be a desire to provide necessary information to the FDA while maintaining the confidentiality of each manufacturer's intellectual property. This can be accomplished by the application holder submitting to the FDA a letter of authorized cross-reference from the owner of the referenced material. This letter would grant the FDA permission to consider the referenced material in its review of the current application. In general, the referenced information may be available from two sources:

1. Existing application: An existing investigational application (investigational new drug [IND] or investigational device exemption [IDE]) or an existing marketing application (new drug application [NDA], BLA, premarket approval application [PMA], or 510(k)) may provide information relevant to a new developer's application. In some instances, the application being cross-referenced may be under co-review for use in the combination product. In other instances, the cross-referenced application may be approved for other purposes, but may have information relevant to the new use.
2. Master file: Master files provide an administrative method to submit confidential information to the FDA when an appropriate

investigational or marketing application for the constituent is not available. A master file is not a substitute for an investigational or marketing application. The FDA neither approves nor disapproves master files; rather, information in a master file is considered in the context of a particular investigational or marketing application. It should be recognized that the information in a master file may be sufficient to support a marketing application for one product, while additional information may be necessary to support its use in another product. For example, this may occur when specific issues raised by the new use of a constituent are not addressed in the master file. Such information could be provided by supplementing the existing master file, or by providing the necessary information in the application under review. More information on drug master files may be found in 21 CFR 314.420 or at http://www.fda.gov/cder/dmf/index.htm. More information on device master files is available at http://www.fda.gov/cdrh/dsma/pmaman/appdxc.html#P7_2.

The information presented above certainly calls for appropriate action on the part of the manufacturer of combination products to not only comply with the regulations but also consistently meet product requirements. We will now present certain key manufacturing steps to be taken by the manufacturers of combination products.

Manufacturing plan

During the development of combination products, it is important to establish a plan that will address all manufacturing-related needs for the product. The plan is usually created such that it is a subset of the overall product design and development plan but provides sufficient details on activities such as process characterization, facility selection and qualification, supplier selection and qualification, activities to improve the necessary competency of personnel, manufacturing risk assessment, validation plan, scale-up, technology transfer, and so on. It should also address any required information technology requirements to support full-scale manufacturing of the combination product. A manufacturing leader supporting the development of the combination product typically develops this plan. This plan is then vetted and validated with many other stakeholders before it is finalized, since this has many financial implications.

Though the manufacturing activities for combination products might resemble typical manufacturing-related activities for a medical device or pharmaceutical or biologics, it is the level of detail and the need to address all the constituents of the combination product that set this plan apart. We will now present additional levels of detail for some of these activities in the manufacturing plan.

Manufacturing risk assessment

Combination product manufacturers, like their counterparts in the medical device, pharmaceutical, or biologics industries, are moving toward the use of a risk-based management approach to various aspects of manufacturing. This includes application of risk management tools and techniques in the design and development of the manufacturing process and facility, validation of IT computerized systems, and so forth.

Combination product process risk management is the effective use of tools and techniques to identify and manage key system, product, and process risks in the development and deployment of the manufacturing process for combination products. This will ensure that there are no product quality issues such as contamination of the final product from upstream steps in the manufacturing processes. The methodology used for the risk assessment can be failure mode and effects analytics (FMEA), failure modes effects and criticality analysis (FMECA), or fault tree analysis (FTA), or combinations of FTA and FMEA. This risk assessment should be performed as a cross-functional effort. The makeup of the manufacturing risk assessment team is based on the nature of the combination product. For example, in the case of a drug-eluting stent, the risk assessment team should include both device and drug manufacturing experts along with product designers, marketing experts, regulatory and clinical specialists, and, if possible, customers. The risk assessment should look into how the product will be manufactured, lessons learned, regulatory response to the risk assessment, and both proactive and reactive approaches to risk assessment.

Regardless of the area of focus for risk management in manufacturing, it is recommended that the combination product manufacturer follow the approach given below:

First, determine the appropriate scope of the risk management effort. This should be followed by a determination of the necessary deliverables, the use and justification of risk-based categories, the importance of GxP-relevant assessments, and, finally, an approach to effectively assess the magnitude or impact of changes within the manufacturing process and facilities.

In our experience, we found that while medical device companies are further ahead in understanding the need for manufacturing risk assessment and taking appropriate steps to manage risks, both pharmaceutical and biologics companies are in the early stages of risk assessment. This difference can play a role when combination product development teams try to create (and manage) manufacturing risk assessment approaches for the product being developed. Teams can often be frustrated with the delay in completing the creation of the risk assessment and might end up blaming the tools, such as FMEA, for their failure.

Process characterization

During the development of combination products, it is important to characterize the manufacturing process. The purpose of process characterization is to challenge the edges of the design space where failures can occur for better understanding of the process. This, in turn, will help in establishing process specification limits after allowing for safety margins. The design team must carefully evaluate process elements and parameters to identify critical parameters that need to be characterized. Documents pertaining to risk analysis, technology, test method evaluation, stability studies, and so on, can play a huge role in this aspect. Of these, the area of technology can become a more critical consideration for combination products. For example, in the case of drug-eluting stents, it is important for the manufacturer to consider the technology that will be used to manufacture the drug-coated stent. In other words, how will the drug be coated onto the stent? Will it be spray coated, dip coated, or coated through other means?

If the combination product includes biologics, significant manufacturing steps that may affect a product's safety, purity, or potency include:

- Inoculation of vessels or animals for production
- Cell culture production and characterization
- Fermentation and harvesting
- Isolation
- Purification
- Physical and chemical modifications
- Required infectious disease testing of blood and blood components
- Blood donor recruitment and maintenance of donor deferral registries

Manufacturing steps important to the purity and integrity of the final product include:

- Chemical and biological testing other than blood infectious disease testing
- Formulation
- Sterile filling
- Lyophilization
- Labeling

If the combination product includes an active product intermediate (API), it is important to understand and implement appropriate pharmaceutical development steps. The aim of pharmaceutical development is to design a quality product and its manufacturing process to consistently deliver the intended performance of the product. Recently, the FDA has proposed the concept of design space. The information and knowledge gained from pharmaceutical development studies and manufacturing experience provide scientific understanding to support the establishment of the design space, specifications, and manufacturing controls. Design space is a multidimensional space that encompasses combinations of product design, manufacturing process design, manufacturing process parameters, and raw material quality that provide assurance of suitable quality and performance. Information from pharmaceutical development studies can be a basis for quality risk management. It is important to recognize that quality cannot be tested into products, that is, quality should be built in by design. Changes in formulation and manufacturing processes during development and life cycle management should be looked upon as opportunities to gain additional knowledge and further support establishment of the design space. Similarly, inclusion of relevant knowledge gained from experiments giving unexpected results can also be useful. Design space is proposed by the applicant and is subject to regulatory assessment and approval. Working within the design space is not considered a change. Movement out of the design space is considered to be a change and would normally initiate a regulatory postapproval change process.

The pharmaceutical development section should describe the knowledge that establishes that the type of dosage form selected and the formulation proposed are suitable for the intended use. This section should include sufficient information in each part to provide an understanding of the development of the drug product and its manufacturing process. Summary tables and graphs are encouraged where they add clarity and facilitate review. At a minimum, those aspects of drug substances, excipients, container closure systems, and manufacturing processes that are critical to product quality should be determined and control strategies justified. Critical formulation attributes and process parameters are generally identified through an assessment of the extent to which their variation can have an impact on the quality of the drug product.

The selection, control, and any improvement of any manufacturing processes intended for commercial production batches should be clearly explained. It is important to consider the critical formulation attributes, together with the available manufacturing process options, in order to address the selection of the manufacturing process and confirm the appropriateness of the components. Appropriateness of the equipment used for the intended products should be discussed. Process development studies should provide the basis for process improvement, process validation, continuous process verification (where applicable), and any process control requirements. Where appropriate, such studies should address microbiological as well as physical and chemical attributes. The knowledge gained from process development studies can be used, as appropriate, to justify the drug product specification. The manufacturing process development program or process improvement program should identify any critical process parameters that should be monitored or controlled (e.g., granulation endpoint) to ensure that the product is of the desired quality.

Significant differences between the manufacturing processes used to produce batches for pivotal clinical trials (safety, efficacy, bioavailability, bioequivalence) or primary stability studies and the final process should be summarized so that the influence of the differences on the performance, manufacturability, and quality of the product is clearly understood. The information should be presented in a way that facilitates comparison of the processes and the corresponding batch analyses information. The information should include (1) the identity (e.g., batch number) and use of the batches produced (e.g., bioequivalence study batch number), (2) the manufacturing site, (3) the batch size, and (4) any significant equipment differences (e.g., different design, operating principle, size).

In order to provide flexibility for future process improvement, when describing the development of the manufacturing process, it is useful to describe measurement systems that allow monitoring of critical attributes or process endpoints. The FDA also proposed the concept of process analytical technology (PAT). This is nothing but a system for designing, analyzing, and controlling manufacturing through timely measurements (i.e., during processing) of critical quality and performance attributes of raw and in-process materials and processes with the goal of ensuring final product quality. Collection of process monitoring data during the development of the manufacturing process can provide useful information to enhance process understanding. The process control strategies that provide process adjustment capabilities to ensure control of all critical attributes should be described.

Characterization of test methods is completed to establish the performance characteristics of them. These characteristics and limitations include environmental conditions, test method stability, training requirements, and use of established compendium test methods. The first step is an assessment of the ability of the process to reliably produce a product of the intended quality (e.g., the performance of the manufacturing process under different operating conditions, at different scales, or with different equipment). An understanding of process robustness can be useful in risk assessment and risk reduction (see ICH Q9, *Quality Risk Management* glossary for definition) and to support future manufacturing and process improvement, especially in conjunction with the use of risk management tools (see ICH Q9).

Design for Six Sigma (DFSS) tools such as design of experiments and measurement systems analysis are extremely useful in process characterization of combination products. Readers are strongly encouraged to refer to the publications listed in the references for further information on DFSS tools.

Facilities, equipment, and media

The selection or establishment of a manufacturing facility is critical in the manufacture of combination products. A traditional medical device manufacturer should not assume that he or she can use the current manufacturing facility to manufacture, for example, a drug-device combination product. If this facility is used to produce a combination product as a single entity or as a copackage, it is safe to assume that both sets of current good manufacturing practice regulations are applicable during and after joining the constituent parts together. If, for example, the drug-eluting cardiovascular stent is manufactured in the same facility, then the manufacturing operations associated with the drug coating or polymer reservoir must comply with the pharmaceutical GMP regulations found in 21 CFR 210 and 21 CFR 211, while the manufacture of the stent and its associated delivery system must comply with the quality system regulation (QSR) found in 21 CFR 820. Looking at a specific situation, for example, if the facility has an electrical switch box in the manufacturing area, it might have been acceptable under the QSR but not under the cGMP regulations, due to the potential of particles that can contaminate the drug constituent.

This understanding should help the manufacturer of a combination product select the right strategy for manufacturing the product. This strategy can be:

1. Select facilities for each constituent part and the combination that are geographically separated.
2. Select facilities for each constituent part and the combination that are physically separated but in the same location.
3. Other appropriate approaches to locate the manufacturing facility.

Manufacturing facilities should establish well-defined areas or other appropriate control systems to prevent contamination or mix-ups for handling and manufacturing combination products. For example, controls should be established so that potent active pharmaceutical ingredients (APIs) are not introduced into the manufacturing environment sampling of APIs. Other examples include having appropriate exhaust systems to prevent possible migration levels of potent compounds in the manufacturing facility. The building design and construction should be adequate to handle manufacturing, processing, packaging, and holding combination products. For example, materials that cannot be easily cleaned in the formulation room (class 10000) and entry/processing room (class 100000) should not be used.

If the combination product includes a biologic constituent, it is important to know that a cell culture is usually done in dedicated equipment, usually made of stainless steel that is connected using many stainless steel pipes. The environment is usually extremely clean, and the equipment must be cleaned, sterilized, and validated before reuse to prevent cross-contamination.

As for the facility requirement for sterilization, further analysis is needed to determine in more detail how, and if, the current extensive infrastructure of large-scale radiation processing facilities can most effectively be used to process combination products. Specifically, the important differences between traditional medical devices and combination products may include:

- Temperature control, that is, irradiation in a frozen state
- Inert atmosphere
- Low sterilization doses with a tight dose uniformity ratio (DUR), which is the ratio of the maximum to minimum absorbed dose in the product (e.g., DUR 1.3/1.0 may be required)
- Small production batches of high-value products amounting to, for example, a monetary value of US$500,000 to US$1 million in a facility
- Just-in-time irradiation
- Additional irradiator controls to further eliminate the possibility of overdosing product
- New strategies to effectively deal with process interruptions

In some cases, this is likely to require modifications of existing irradiators to accommodate the special conditions associated with the sterilization of drug-device combination products. Critical for success is working with the irradiation service provider early in the product development process.

In the *Guide to Inspections of Microbiological Pharmaceutical Quality Control Laboratories*, the FDA provides the following guidelines regarding facilities: Begin the inspection with a review of analyses being conducted and inspect the plates and tubes of media being incubated (caution should be exercised not to inadvertently contaminate plates or tubes of media on test). Be particularly alert for retests that have not been documented and "special projects" in which investigations of contamination problems have been identified. This can be evaluated by reviewing the ongoing analyses (product or environmental) for positive test results. Request to review the previous day's plates and media, if available, and compare your observations to the recorded entries in the logs. Inspect the autoclaves used for the sterilization of media. Autoclaves may lack the ability to displace steam with sterile filtered air. For sealed bottles of media, this would not present a problem. However, for nonsealed bottles or flasks of media, nonsterile air has led to the contamination of media. In addition, autoclaving less than the required time will also allow media-associated contaminants to grow and cause a false positive result. These problems may be more prevalent in laboratories with a heavy workload. Check the temperature of the autoclave since overheating can denature and even char necessary nutrients. This allows for a less-than-optimal recovery of already stressed microorganisms. The obvious problem with potential false positives is the inability to differentiate between inadvertent media contamination and true contamination directly associated with the sample tested. The United States Pharmacopeia (USP) points out that the facilities used to conduct sterility tests should be similar to those used for manufacturing the product. The USP states: "The facility for sterility testing should be such as to offer no greater a microbial challenge to the articles being tested than that of an aseptic processing production facility." Proper design would therefore include a gowning area and pass-through airlock. Environmental monitoring and gowning should be equivalent to that used for manufacturing the product.

Contract manufacturing and testing

With the industry focus on cost containment, use of contract manufacturers and test facilities is increasingly becoming the norm rather than an exception. There are also contract packers or labelers, custom grinders, and so on. All these extramural facilities are typically considered an extension of the manufacturer's own facility. When this is the case, we strongly encourage the combination product license manufacturer (or the "owner" of the approved combination product) to carefully ensure that the contractor will meet all necessary regulations to manufacture and test the product. Because the contract manufacturers are engaged in the manufacture of a drug or device, they must comply with applicable provisions of the *Food, Drug, and Cosmetic Act* (*FD&C Act*) (21 USC 301 et. seq.) and applicable regulations.

In the next few paragraphs we will focus on a biologics contract manufacturing scenario. Please note that the logic and thought process is applicable to other constituent part manufacturers as well. If the contract facilities manufacture biologics, these facilities are subject to FDA inspection under section 351(c) of the Public Health Service (PHS) Act and section 704(a) of the *FD&C Act*. Facilities performing contract operations for biological products must register with the FDA in accordance with registration and listing provisions in 21 CFR Parts 207, 607, and 807. Because the combination product license manufacturer must ensure that the contract site complies with applicable product and establishment standards, the license manufacturer should have access to floor plans, equipment validation, and other production information. The license manufacturer must have a procedure in place for receiving information from the contract facility on all deviations, complaints, and adverse events (21 CFR 600.14(a), 606.171(a), 803.10). The contract manufacturer should fully inform the license manufacturer of the results of all tests and investigations regarding or possibly having an impact on the product. The FDA expects the license manufacturer to assume responsibility for compliance with the applicable product and establishment standards (21 CFR 600.3(t)). Therefore, if the license manufacturer enters into an agreement with a contract manufacturing facility, the license manufacturer must ensure that the facility complies with the applicable standards. An agreement between a license manufacturer and a contract manufacturing facility normally includes procedures to regularly assess the contract manufacturing facility's compliance. These procedures may include, but are not limited to, review of records and manufacturing deviations and defects, and periodic audits.

The license manufacturer should be aware that all contract manufacturing locations must be in compliance with appropriate regulations and are subject to inspection from the time of the submission. It is important that the license manufacturer obtain documented assurance from the contractor that any FDA list of inspectional observations will be shared with the license manufacturer to allow evaluation of its impact on the purity, potency, safety, and so on, of the license manufacturer's product. In accordance with 21 CFR 200.10, the FDA may disclose any information obtained during the inspection of a contract facility having a specific bearing on compliance with the *FD&C Act* to the license manufacturer.

For each contract arrangement, the license manufacturer's BLA/supplement should describe the product subject to contract manufacturing, including:

- Product stability and the manner of shipment to and from the contract facility
- Responsibilities of each participating entity
- Contract manufacturer's name, address, license number, if applicable, and registration number
- List of all standard operating procedures applicable to the contract arrangement

Contract firms that do not wish to provide all necessary information to the license manufacturer should consider a shared manufacturing arrangement.

Similar to using contract manufacturers for manufacturing combination products, many manufacturers contract with private or independent testing laboratories to analyze their products. Since these laboratories will conduct only the tests that the manufacturer requests, it is important to determine the specific instructions to be given to the contractor. These instructions must then be evaluated to ensure that necessary testing will be completed. For example, in the case of a drug-coated stent, if the manufacturer enlisted the support of a contract testing service to evaluate the coating thickness, it must make sure that the contract manufacturer has the necessary validated test equipment to measure thickness consistently. Analytical results, particularly for those articles in which additional or retesting is conducted, should be reviewed. Test reports should be provided to the manufacturer for tests conducted. It is not unusual to see contract laboratories fail to provide complete results, with both failing and passing results. Bacteriostasis/fungiostasis testing must be performed by either the contract lab or the manufacturer. These test results must be negative; otherwise, any sterility test results obtained by the contractor on the product may not be valid.

Sterilization of combination products

Regulators, when comparing aseptic processing with terminal sterilization, are saying that wherever possible, terminal sterilization is preferred. This is because it lowers the chance of error and the risk of a contaminated product causing infection or transmitting a disease to the patient. However, sterilization of combination products can be problematic, due to drug/biologic and sterilant incompatibilities. Radiation and ethylene oxide sterilization cycles may degrade drug/biologic characterization and potency. Impurities may actually increase due to these harsh process parameters. Aseptic processing, for instance, coating drug products, can be used as an alternative to final product sterilization through radiation or ethylene oxide gas. A dry-heat sterilization process can be an option for most small-molecule combination products.

There are several areas that require further attention and these are discussed below.

Lower sterilization dose

Because the efficacy of drug or biologic molecules is likely to be dose dependent, every effort should be made to determine the lowest possible radiation dose that is required to make the products safe for their intended use. The current dose-setting guidelines of the International Organization of Standardization and the Association for the Advancement of Medical Instrumentation have a number of limitations when they are applied to these types of products. They do not work well for ultra-ultraclean products, which will often be the case with drug components with a bioburden of less than 0.1 colony-forming unit per product unit.

Dose setting using VDmax can allow a Dmin of 15 kGy, but if products have already been aseptically filled or are ultraclean, a dose in the order of 7 kGy may be all that is required. Current dose-setting methods require the sacrifice of a large number of product samples for dose setting or verification dose testing; the high cost and small volume batch size of combination products may make this prohibitive.

The definition of sample item proportion (SIP) and selecting the SIP need to be reexamined because drugs can be in bulk powder form in relatively large packages and worth thousands of dollars/euros per gram. It is questionable whether the standard bioburden population and distribution used in the current dose-setting methods for medical devices are applicable to drug (fluid) products. The establishment of sterilizing doses for medical devices using current dose-setting methodology takes into consideration product bioburden and its resistance to radiation, but is currently tied to an arbitrary sterility assurance level (SAL) of 10-6. The medical device and pharmaceutical industries and their regulators may define sterility in different terms.

A collaborative approach effectively resulted in the development of the international standard ISO 11137, Parts 1, 2, and 3:2006, *Sterilization of Health Care Products—Radiation.* To address the issues associated with combination drug-device products, industry experts and various regulatory authorities will need to work together and reconsider the product SAL that is essential for patient safety, prescribed use of the product, evolving microbial challenges, and the benefits to healthcare and patients.

Dosimetry and modeling

Because it is incumbent on us to minimize the sterilization dose, it is of increasing importance to select the dosimetry system that has the minimum inherent number of uncertainties or complications associated with its use. It should cover the broadest dose range and ideally experience no dose rate effect. Dosimetry for refrigerated products may need a temperature correction factor and be calibrated accordingly. Therefore, alanine, which is of reference standard quality and performs in a dose range of 0.02–200 KGy, may become the dosimeter of choice for these drug-device combination products.

Radiation processing potentially has a strong role to play in the terminal sterilization of advanced combination drug-device and biologically based products. Looking ahead, advances in the following areas can be envisioned:

- Lowering of the sterilization dose to reflect intended use and product need
- Increased use of more sophisticated dosimetry and modeling
- Improved process control in existing facilities and the design of new systems
- More irradiation of product in the frozen state
- Development of additional free radical scavengers
- New radiation-compatible materials and innovative packaging
- Development of applicable international standards and regulatory harmonization

Packaging and labeling

All healthcare products are typically packaged and remain packaged until they are used. However, there are differences between medical device packaging and drugs/biologic product packaging. While a medical device package protects the product, allows for sterilization, and maintains its sterility until use, drugs and biologic product packages focus on the safety and efficacy of the product until its expiration date.

Other features of medical device, drug, and biologic packaging include:

- Medical device packaging materials must be nontoxic and biocompatible, with porous materials providing adequate microbial barrier. In addition, sterilization should not affect materials or packages. Further, aging or stability studies are typically used to demonstrate the adequacy of the physical properties to protect the device from damage and maintain sterile package integrity until its expiry date.

- Drug packaging follows the FDA's guideline for container closure systems for packaging human drugs and biologics, which was published in May 1999. Drug packaging should focus on using suitable packaging components, batch-to-batch uniformity of these components, the distribution system to be used, use of resign that is already on FDA/EU status, use of suppliers that are vertically integrated, and so on. Chapter 661 in USP should also be used when developing packaging for drugs.
- "In contrast to most drugs that are chemically synthesized and their structure is known, most biologics are complex mixtures that are not easily identified or characterized," explains copy on the CBER website. "Biological products tend to be heat-sensitive and susceptible to microbial contamination. Therefore, it is necessary to use aseptic principles from initial manufacturing steps, which is also in contrast to most conventional drugs. Biologic products often represent the cutting-edge of biomedical research and, in time, may offer the most effective means to treat a variety of medical illnesses and conditions that currently have no other treatments available."

With combination products, packaging has to tackle the difficulties associated with delivering a sterile device that now includes a biologic/drug. New materials with maximum barrier properties are often necessary. In some cases, such as transdermal patches or insulin injector pens, the device itself becomes a package for the drug, leading to complexity in shelf-life demonstration, testing, labeling, and so forth. In other cases, combination products include devices and drugs that are copackaged. In the case of a drug-coated stent, packaging materials should be carefully chosen to allow sterilization but at the same time not have an impact on the safety and effectiveness of the drug. This requires a very different focus and mindset from the traditional device- or drug-only packaging approach. Regardless of the configuration, the manufacturing, shipping, and handling issues for combination products are significant. As with the case for the design of combination products, it is recommended that the manufacturer follow a hybrid system for the packaging design and manufacture.

Storage, logistics, and shipping

This is one of the key areas that is often overlooked in the manufacture of combination products. This could be due to lack of awareness of requirements or the assumption that the current controls are adequate. Storage issues such as manufacturing raw materials storage, in-process product storage, and finished product storage should be adequately addressed for all constituents of the combination product. Adequate provisions must be provided for quarantine/released product. Stability issues that can arise when a product is scrapped/wasted, stored, and shipped, as well as the ones used for quality testing, should be addressed. In the case of a drug-coated stent, storage, shipping, and handling issues that can potentially impact the stability of the drug should be addressed up front. If a contract test facility is used, one should adequately address how the coated-stent test samples will be stored, shipped, and handled from/to the contract test facility.

Process and method validation

Process and method validation of combination products is unique because manufacturers must meet two sets of quality system requirements. While the quality system regulation (QSR) requirements apply to medical device manufacturing, current good manufacturing practices (cGMP) requirements apply to drug and biologic manufacturing. For example, in the case of a drug-coated stent, validation of the processes used to manufacture device and drug constituents must be completed before the combination product can be marketed. While the process to manufacture the stent and the delivery system follows the QSR requirements, the process to manufacture the drug follows cGMP requirements. In the case of a process such as stent coating with the drug, the validation should be performed to demonstrate that the process can consistently manufacture a coated stent that consistently meets established targets. In this instance, the debate between which system should be used becomes a secondary item as long as the validation is performed as stated above. As far as the test/analytical methods are concerned for the drug-coated stent, they must be validated prior to process validation to ensure that the decisions made are based on true process output rather than one that includes method inaccuracies.

Process validation includes many different activities, such as:

- Writing process validation protocols per established timelines and ensuring buy-in from all stakeholders regarding scope, approach, technical content, accuracy, data analysis approach, and so on.
- Executing process validation protocols following established timelines or ensuring proper execution of protocols.
- Writing process validation completion reports that are accurate and compliant with internal/customer requirements and regulatory expectations.
- Resolving any discrepancies, deviations, or departures.
- Performing revalidation and partial validations as appropriate, and determining specific distinctions related to computer systems validation.
- Training employees on all aspects of process validation following established training guidelines.
- Effectively deploying DFSS tools such as design of experiments (DOE) or failure mode effect analysis (FMEA) in validation to challenge the process and identify which factors to control. Other DFSS tools, such as

statistical process control (SPC), should also be utilized to show that the process is stable, can produce consistent outputs over time, and meets specifications with a high degree of confidence.

In the case of combination products that include a biologic constituent, it must be noted that due to the stringent requirement associated with validation, manufacturers of biologic products have historically opted for the use of dedicated lines for each product line. This leads to low equipment utilization rates, increasing costs. However, recent improvements in technology have resulted in cheaper single-use equipment, which reduces the need for cleaning, cleaning validation, and so on.

On the pharmaceutical side, recent FDA guidances on process analytical technologies (PATs) and quality by design (QbD) on process validation must be evaluated. Risk-based process validation and the application of DFSS tools such as FMEA are also introducing changes in how process validation is performed. Note that these approaches are not new to the medical device companies due to the introduction of the FDA's design control and Global Harmonization Task Force's (GHTF) process validation guidelines.

Regardless of the quality system used, it must be noted that not all process validations for combination products need to be completed before initiating a clinical trial. We encourage the reader to have an ongoing dialogue with the regulatory agencies on the process validation approach for combination products.

In this chapter, we gave an overview of the manufacturing aspects of combination products. Where possible, we also provided some details using the drug-coated stent as an example. We acknowledge that there is more information available to cover the topic of manufacture of combination products, and the reader is encouraged to seek out knowledge and expertise regarding the manufacture of combination products.

Bibliography

Butschli, Jim. Combination products challenge packaging. *Packaging World,* June 2006.

Guide to inspections of Microbiological pharmaceutical quality control laboratories. FDA, July 1993.

Masefield, J., Brinston, R. Radiation sterilization of advanced drug-device combination products. *Medical Device Technology,* March/April 2007.

Matlis, Daniel R. Making life-sciences from life. White paper prepared for CAMSTAR, June 2007.

Triolo, Phil. Coated Combination Products: Regulation and Technology. *Medical Device & Diagnostic Industry,* January 2005.

chapter seven

Challenges and pitfalls to avoid with combination products

In light of medical device, pharmaceutical, or biotech companies joining forces to codevelop combination products, creative and innovative technologies are emerging to address the various medical needs of current patient populations. As mentioned in Chapter 1, these combination products can provide multiple and complementary modes of action, which may result in greater therapeutic benefits than the drug, biologic, or device working alone. In many instances, the combination products may enable diseases to be treated with localized therapeutic effects, such as a device delivery system providing a localized drug release at a particular disease site in cases where systemic treatments may pose greater risk and harm to patients. We provided an overview of challenges in combination product design, development, manufacturing, regulatory, and so on, in the previous chapters. In this chapter we will present an in-depth look at the challenges and pitfalls to avoid with combination products.

Challenges and pitfalls

Combination products pose significant development, regulatory, business, and technical challenges for those companies that are contemplating entering the combination product arena. The combination of a drug, device, or biologic product would require not only the merging and understanding of the technical fields, but also the bringing together of the knowledge base from the engineering, biologic, and pharmaceutical perspectives in terms of product development, quality, regulatory, and related support systems needed to develop a successful combination product.

A pharmaceutical organization that looks at developing a combination product with a device company would need to address the new required technical interfaces and supporting infrastructure for the development team to develop the product in a hybrid environment. Because manufacturers of combination products have to integrate new and different technologies, the challenges faced by these companies include new ways of integrating the device/drug or biologic/device development methods, regulatory requirements, competency in key individuals heading the projects, and so on.

It is important to revisit the classification of combination products at this stage. As defined in Chapter 2, the classification of combination products can be generally categorized into four separate categories based on their specific components: novel drug delivery systems, traditional drug delivery systems, drug-enhanced devices, and regenerative medicinal products.

1. *Novel drug delivery systems*: Examples in this category include patches, transdermal or intradermal injections, inhalation devices, sprays, and drug-eluting disks, which typically combine existing drugs with *new delivery devices.* These products are designed to improve the convenience and comfort of administration, and drug effectiveness through localized administration, and to enable delivery of a drug through a different administrative route other than oral, subcutaneous, or intramuscular injections. In many cases, these changes to administration, or drug formulations and bioavailability could potentially increase the drug development technological challenges, although these products have a moderate complexity. These products would be primarily governed by the Center for Drug Evaluation and Research (CDER) through its regulatory pathway for drugs since the primary therapeutic mode of action would be drug related in these instances.
2. *Traditional drug delivery systems*: These products combine or package drugs together with injection devices to improve convenience of administration. This would include pen-based delivery systems, drug pumps, profiled syringes, and autoinjectors. Since the drug-device interface is relatively simple in these products, the components in this instance can be developed separately and integrated in the later stages of the product development cycles. These components can also be regulated separately using the established regulatory regimes for drugs and devices.
3. *Drug-enhanced devices*: Products such as drug-eluting stents, bone cements with antimicrobial agents, coated catheters, other devices with antimicrobial coatings, and infective sutures can enhance the functionality, performance, or efficacy of these devices.
   In large instances, these products combine existing devices with existing drugs. In these cases, the drug-device interface is often novel and is oftentimes critical to the performance of the combination product. As a result, development of these products is more complicated than similar device-only products. In these instances, since the primary therapeutic action results from devices (CDRH), device regulators would primarily govern these products, with a secondary review from drug-related regulatory agencies (CDER).
4. *Regenerative medicinal products*: These are products that combine devices with biologically active substances to facilitate/enhance healing and regeneration of damaged tissues. The device often serves as the support structure for the growth of the biologic component with the product often being an implant. Examples of such products include absorbable

meshes for bone growth, Dermagraft (human fibroblast–derived dermal substitute), artificial replacement organs (such as the bioartificial pancreas), coated spinal fusion cages with recombinant human bone morphogenic proteins, and so on. These are among the most complex combination products, as they have to consider the interaction between the product and the body's physiological response to the product. As a result, the development process for such products is also extremely complicated and integrated, because the components under development have to be tightly coupled. The lead review and oversight by the relevant agency would vary in these cases, as the primary mode of action for these combination products may vary on the specific combinations.

Management responsibilities

Key areas facing management include determining a successful regulatory strategy; providing strong commitment and resources to the combination product program; providing a culture that focuses on quality first; establishing policies, manuals, and audit mechanisms to assess and ensure the compliance of the combination product development; and providing adequate personnel with the necessary skill sets, education, background, training; and experience.

In the case of a drug-coated stent, the key organizational challenges that companies face stem from the need to adopt an integrated mindset that fuses design practices from the device world (stent and delivery system) with the process-focused thinking from drug development (formulation of drug product). To keep development efforts from stalling, companies need to identify the right skill sets, capabilities, partnerships, and alliances. It is important to integrate the relevant aspects of the device and pharmaceutical development, recognizing that different methodologies, terminology, and backgrounds of the staff are involved. Companies must establish the proper project governance, team structures, and functional involvement needed for combination product development and launch to succeed in the market.

In addition, companies need to correctly estimate the financial costs and timelines involved in the development, manufacture, and meeting of the compliance requirements for combination products. To minimize the risks mentioned above, companies have to evaluate organizational capabilities and development practices both internally and in terms of what a tentative partner could bring to the table. In many situations, the product development effort can be used to restructure existing organizational structures, product development processes, and roles and responsibilities of the key team members. This can result in a novel system that is not only effective but efficient as well in bringing combination products to the market for the particular company successfully. Leadership that is not mature enough to deal with these challenges might either cancel the combination product programs and resort to their traditional strengths (development of drug or device, etc.) or shy away from investing in these technologies.

Collaborative partnerships

When medical device, biologics, or pharmaceutical companies enter into partnerships and collaborative arrangements with other companies or manufacturers, there are numerous challenges that result from changes and differences in management systems, quality, regulatory, product development, financial, logistic issues, and so on. These issues are further heightened when the partners are from different worlds, for example, biotech merging with a device, or a device company partnering with a pharma company, or a device partnering with a biologic company.

Drug and device companies, while sharing a common regulatory agency, some similar regulatory constraints, and even some overlapping regulatory vocabulary (e.g., registration and misbranded), may enter into negotiations with fundamentally different needs, perspectives, and expectations. These differences need to be kept in mind when device companies begin talking with potential pharmaceutical or biologic partners, when negotiating an agreement, and once the companies commence their collaboration. If not, the consequences include loss of trust, credibility, company finances, and the desired outcome.

Examples of partnership requirements include:

1. A device company about to partner with a drug company must appreciate from the start that the two companies would have different perspectives.
   - One of the principal issues that need to be addressed for a drug-device collaborative project is determining the key FDA center responsible for regulating the product. Since devices and drugs are regulated so differently, this question needs to be resolved at an early stage of the partnership.
   - In many cases, the regulatory framework for a combination product is established by precedent. Drug-eluting stents are regulated as devices, while a prefilled syringe will be regulated as a drug. Additionally, an antibiotic-coated orthopedic implant may be regulated as a device if the intended use of the antibiotic is to prevent colonization on the implant. The same product with a claim that the coating prevents infection might be classified as a drug.
2. If a U.S.-based device company is partnering with a foreign drug-biologic company, both sides need to realize not only differences in the technology and regulatory requirements, but also significant cultural differences.

Developmental challenges

Development team

The key challenges in development for combination products are how to address the scientific and technical issues when combining device, drug, or biological product constituents. For example, if a predominantly pharmaceutical company is developing a drug-device product, it might not know that performing plastic mold-flow analysis can improve the performance of the plastic parts used in the product. During the development of innovative combination products, additional challenges faced by the development teams are regarding the regulatory strategy around the combination product to ensure all requirements are met for the multiple, regulatory agencies. In many instances this is a difficult area to address, as the regulations, marketing applications, quality compliance requirements, and postmarketing regulatory requirements are significantly different based on the precise combination product.

Defining the type of preclinical or animal studies that are appropriate prior to conducting human studies can be challenging to the development team. Defining the requirements for the design of the clinical trials, and so on, is an additional area of discussion for the product development team. Other challenges include identifying the critical path to progress from concept to development to product commercialization, and the timelines that need to be defined realistically to be successful.

For *device companies* that have to design combination products, collection of requirements and development of specifications for components may include engineered devices, drugs, or biologics with pharmacological, immunological, or metabolic mechanisms of action. In addition to designing better products, the manufacturers need to perform increasingly robust and thorough testing. They must also define the types of changes that can be made to the commercial manufacturing process with an understanding of the effect on the product, its sterility, and its stability.

The product development team and supporting functions therefore need to develop a successful strategy, common development process that is understood and executed by the cross-functional team, and the correct mix of technical, business, regulatory, clinical, and quality team members capable of supporting the project adequately.

For example, product development teams for drug-enhanced devices have one of the biggest design and manufacturing challenges in ensuring adequate sterility and shelf life of the drug on a device. Coatings can flake off, coatings can cover the stent, or the product can deposit too much drug into the body, all of which could reduce efficacy and create a safety risk for end users. To manage such problems, manufacturers must perform more product and process characterization earlier in the development cycle than they would with a traditional device. These processes have to be supported by robust analytical and physical methods to establish a fundamental understanding of the relationship between process parameters, design inputs, and product stability.

Additionally, a single approved device and an approved drug may work fine as stand-alone products but pose challenges in terms of technical and scientific functionality when combined. New and creative processes need to be developed for evaluation of safety in preclinical/animal studies and, additionally, effective design of the clinical trial to establish safety and effectiveness of the combined product.

A key consideration for a development team when adding a drug or biologic to a device is the development of test methods for the combination product. The skills and infrastructure required for these different types of test methods are significantly different and should be appropriately addressed by the development team for the combination product:

- Generally, medical device test methods often focus on physical and mechanical properties, whereas pharmaceutical/biologics test methods focus on analytical, bioanalytical chemistry, and biological potency assays.
- In pharmaceutical/biologics, microbiological tests are typically validated and executed by the sites. Analytical methods developed in R&D must be eventually developed into commercial test methods.
- Transfer of complex, analytical chemistry test methods is more familiar in pharmaceutical/biologics companies than in their device company counterparts.
- Device companies should expect a higher scrutiny regarding the pharmaceutical/biological test methods from the regulatory agencies.

Technical challenges

During product development, the project team often looks to the regulatory and quality professionals to provide guidance to the interdisciplinary project teams. Questions are typically posed looking for suggestions and strategies on the preclinical and clinical testing, product characterization and validation of the manufacturing processes, sterility and packaging requirements for the novel combination product, stability requirement if a biologic or drug component is part of the combined product, and so on.

Some of the technical challenges faced by product teams include:

1. Establishing release criteria for the in-process and final product that are reflective of the actual product performance, and to develop supporting characterization assays that can be used to assess the mode of action for the product for manufacturing consistency.
2. Defining preclinical requirements and objective clinical endpoints to assess product performance in regards to safety and effectiveness and developing valuable study designs. Use of third-party evaluation for assessment of safety during the study as new data are gathered.
3. Assessing adequate storage and transport methods for delivering the product to the end user, ensuring that the product integrity, sterility, and performance characteristics are not compromised.

Communication is critical in overcoming these challenges during development both internally within an organization and by its partners, but also with regulatory bodies to ensure that key issues and concerns are discussed with the agency during the development of the combination product and not left until the end.

As new products are developed by organizations, the team has to be aware of any changes that may occur to the regulatory requirements during development.

Yet, for these relationships to succeed, device companies must appreciate and navigate many hurdles unique to partnerships with pharmaceutical or biologic companies. Only by identifying and addressing some of the associated regulatory and commercial issues can manufacturers successfully create combination products.

One major difference between the drugs and devices is that drugs take much longer to reach the market. From the time a drug enters preclinical studies to the time the FDA approves a new drug application (NDA), 10 years or more can elapse. Drug companies are accustomed to such long-term planning; device companies are not. Even if the collaboration begins during a drug's phase 2 trials, it may be years before the drug product is approved. Devices can go through several generations in the time a single drug moves from research and development (R&D) to approval.

A different consideration is that the probability of success for drugs is much lower than that of devices. According to the FDA's statistics, only about 8% of drugs that start phase 1 trials actually get approved, and 1 out of 1,000 compounds that enter preclinical testing reaches the clinical study stage. Although the FDA's Critical Path Initiative and new phase 0 option (for early clinical testing with low doses) are intended to improve the odds, it will take years for the benefits of these programs to be felt.

Not all devices that are developed also make it through the FDA process; however, drug companies face far more daunting odds. Therefore, in early-stage collaborations with drug companies, device manufacturers must take into account longer timelines and lower probabilities of success for their partner's product. If the drug is already marketed and only an NDA supplement is needed (e.g., approval for a new intended use), these differential factors diminish, but they do not vanish.

A drug development program for pharmaceutical companies is also more costly. It is estimated that pharmaceutical companies on average spend more than US$800 million for each approved drug, which includes the costs of failed studies. There is no disputing that the drug development process is extremely costly, far more so than for any device. Drug companies, as a result, face different financial structures. Drug companies of late have focused on so-called blockbusters, that is, where the sales of these drugs exceed US$1 billion annually. Many drugs have hit that target; however, few devices enjoy those kinds of magnificent sales. Even drugs that may be considered poor sellers may exceed US$100 million in sales, a more significant level than the vast majority of devices. For small biotechnology and drug companies, the economic considerations and returns on investment can be very different from those for a device partner, resulting in very different needs and expectations from both partners.

Yet another key difference is that drugs can remain on the market for decades unchanged upon approval, whereas rapidly advancing technology and continuous product modifications typically characterize the device industry. Older devices often get obsolete; however, older drugs can continue to maintain healthy profit margins. The different regulatory requirements and constraints do not allow drug companies to easily make changes due to stringent requirements for altering product or labeling, while devices can quite easily modify product or labeling via a 510(k).

In reference to the quality systems, both collaborative partners are subject to good manufacturing practices. In discussion of GMP, a drug company may state that its device partner is in full compliance with GMP when in reality some drug GMP requirements (such as stability testing, analytical tests, returned product, environmentally controlled environments, etc.) are missing. Conversely, the drug GMP regulation does not have elements such as design controls, managements systems, and corrective and preventive actions normally used by device companies. The physical structure and processes of a device manufacturing facility may be radically different than those of a drug company.

Key questions collaborators need to address are listed below:

- What is the nature of the combination product?
- Are the indications and intended use for the product defined?
- What company, the drug or the device partner, controls the regulatory process?
- Do the collaborators agree on the primary FDA jurisdictional division assigned to the combination product?

Preclinical/clinical studies

During the development of innovative combination products, some of the challenges faced by the project teams are what type of preclinical or animal studies are appropriate prior to conducting human studies. What are the requirements for the design of the clinical trial? Other challenges included by the development and design team are also around the critical path to progress from concept to development to manufacturing the product prior to launch.

For example, a single approved device and an approved drug may work fine as stand-alone but pose challenges in terms of technical and scientific functionality when combined. New and creative processes need to be developed for evaluation of safety in preclinicals/animal studies and, additionally, effective design of the clinical trial to establish safety and effectiveness of the combined product. The development of an appropriate preclinical testing program is a critical step on the path to market availability of these novel products.

Preclinical challenges

It would be interesting to see how CDER and Center for Biologics Evaluation and Research (CBER) view preclinical safety testing of small-molecule drugs and biologics, and how the differences translate to the requirements of both agencies. For example, CDER typically requires toxicology studies in at least two animal species (one rodent and one nonrodent) prior to human testing, while CBER often requires data from only one species. In addition, CDER prefers that first-in-human studies testing synthetic drugs be conducted in healthy volunteers, while CBER typically encourages sponsors to conduct their initial clinical studies in diseased patients.

Another interesting challenge for both sponsors and the FDA will be to determine what absolute amount of clinical trial data that supports combination product safety and efficacy will be required for marketing approval. For example, will CDER require that two consecutive phase 3 clinical trials be performed for device and small-molecule drug combination products, as they typically do for conventional drugs? The FDA's stance on this issue will probably be determined on a case-by-case basis, influenced by any of the following factors: What primary therapeutic claims are being made for the product? Is the presence of the drug on the device simply to enhance a structural function for the device? Is the drug permanently attached to the device, thereby exerting only a local effect, or is it designed to leech off the device into the local environment?

If you are adding a drug product to a device in the development of a combination product, the following steps need to be performed.

Preclinical in vivo *studies*

There are many contract research organizations (CROs) located globally with extensive experience in performing good laboratory practice (GLP)-compliant preclinical safety studies on either medical devices (meeting ISO 10993 standards) or drug-biologics (meeting ICH S1–S7 guidances). However, the challenge in many instances is to locate testing facilities with the appropriate mix of technical and surgical expertise, animal models, and analytical methods to allow a rigorous assessment of the relative contributions of each component of the combination product toward any toxicities associated with its use.

Manufacturers are faced with a significant challenge when choosing appropriate animal models and protocol designs for safety testing of combination products. They must perform rigorous toxicity testing of the combination product, using it intended in the animal model. However, these manufacturers also need to allow concurrent assessment of many of the standard safety parameters required for the preclinical evaluation of the stand-alone constituent of the combination product, whether it is medical devices or drugs and biologics.

The manufacturer is required to conduct animal studies, including feasibility and GLP studies to support research, regulatory submissions, and design validation, including clinical studies at various phases of the new product development process.

The pharmaceutical steps required to be conducted by the device company would include the following:

Early pharmaceutical kinetics (PK) and metabolism studies:

- Medical device companies incorporating a pharmaceutical constituent need to determine local/target organ concentration and PK/safety profile.
- Be aware that existing information may not be sufficient to support the intended application.
- If systemic circulation is required for the combination efficacy, a methodology to determine levels may be needed.
- Companies need to determine that systemic exposure is adequate for efficacy but not toxic.

In vitro metabolism:

- *In vitro* metabolism studies are one of the many approaches utilized to obtain predictive information on the bioavailability, pharmacokinetic profile, and efficacy of a potential drug. As such, it is important that device companies developing combination drug-device products ensure that the correct skill sets reside within the project team so that the external contract research organization (CRO) conducting this activity can be appropriately supervised.
- *In vitro* metabolism information is required. If this information exists for an already marketed pharmaceutical, it can be utilized when applying the pharmaceutical to a device.

Safety pharmacology studies:

- Safety pharmacology studies are required when a drug is applied systemically.
- Device companies may need to understand pharmaceutical impacts in locally applied areaa (e.g., the drug on a stent).

Absorption, distribution, metabolism, and excretion (ADME) studies:

- Medical device companies can utilize previously existing data for the pharmaceutical constituent of the combination product.
- Medical device companies must determine whether any tissue-specific metabolism is occurring locally.

Reproductive toxicology:

- If a medical device company uses a pharmaceutical, this information is required. Existing information can be utilized. Additional reproductive toxicology studies may be required if the local administration provides dose concentrations higher than previously studied.

- In a medical device company, toxicology study species are often chosen based on anatomical similarity. In the pharmaceutical companies, one rodent and one nonrodent species are required for toxicity.

Mechanistic toxicology:

- Mechanistic toxicology is used to reduce risk.
- When the pharmacological evaluation of the compound is made, if at this time you see any adverse reaction to the pharmaceutical element, these studies can be applied to understand rationale.
- If an approved drug is used, mechanistic toxicology is not required systemically. Device companies need to determine if local concentration exceeds concentrations studied.
- Special toxicology studies may be required depending on organ impact, and so on.

Carcinogenicity studies:

- These studies assume systemic delivery of the drug. Only rats and mice are acceptable species for carcinogenicity studies.
- A medical device company that uses a pharmaceutical must understand the toxicology of that pharmaceutical constituent.
- For medical device companies using pharmaceutical elements with local administration, these studies may be required. Existing study results can be utilized.
- If local concentrations are higher due to local delivery via device, studies may need to be reconducted. Medical device companies also need to consider the impurities that come from combination products and ensure that these are studied.
- If local delivery is continued for sustained periods, then a systemic absorption profile may require carcinogenicity studies.

Preclinical testing

For combination products, it is necessary that the manufacturer be able to characterize the actual product intended for clinical study so that regulatory agencies can know how the results for the clinical trial apply to the commercially distributed product.

In cases where a company is developing a combination product, comprising a drug or biologic component in combination with a device that has already received approval for use in humans, it may be possible to build efficiencies and leverage information from the approved device to some extent into the preclinical safety testing program with prior discussions and agreement from the FDA.

In these cases any previous data on the actual toxicities of the biologic or drug when administered as a single agent (not in association with a device) to humans should be included, or cross-referenced, in submissions to the FDA, and potentially reevaluated in additional studies.

Even though a combination product may include a previously approved biologic or device, it is likely that the intrinsic properties of the drug or biologic could be altered by physical or chemical attachment to a medical device. As an example, if the half-life of the drug will be increased by its attachment to a device, it is possible that either the nature or the time course of the toxicities due to the drug could be altered. In this case, it may be useful to consider the attached drug as a new chemical; the attached molecule should be considered a new chemical entity during the design of the initial preclinical studies.

An interesting challenge in the development of combination products is the potential change in immunogenicity of a drug or biologic when it is attached to the surface of a device surface. For example, when recombinant proteins are bound to a device surface, it may be possible that these molecules will have altered domains; consequently, they may be recognized by the immune system of the host as foreign. In cases where the recombinant protein is identical or very similar in structure to a naturally occurring protein in the body, an actual autoimmune response against the natural protein can potentially occur.

However, since most device-drug combination products involve the attachment of drugs that are small-molecule synthetic drugs to a device, it is usually presumed that host immune responses against these molecules are not an issue. Once these small-molecule drugs bind to the surface of medical devices, they become multivalent and are recognized by antibody-producing B-cells in the host.

Since the majority of implanted medical devices are in many cases associated with minor local inflammation, it is possible that small-molecule drugs and biologics that would otherwise have very little to no direct interaction with immune cells in the host could directly interact with these same cells when bound to the surface of a device. Drugs or biologics coated onto medical devices could potentially affect circulation patterns or release of soluble factors from various immune cells. This, in rare cases, may have the potential to alter the functioning of the immune system, either locally at the site of device implantation or systemically.

In regards to preclinical testing, the FDA conveys that a single-most-critical message in the characterization of the finished sterilized product as it is to be studied is essential. It is certainly understood by the FDA that changes and design improvements will normally occur during the product development process; however, during the time the company is ready to perform its clinical trial, the company needs to have characterized the actual product intended for the clinical study so that the FDA can evaluate the clinical trial results and how they apply to the commercially distributed product.

Taking the drug-eluting stent as an example, some of the most critical areas include characterization of the coating and the drug substance, *in vitro* and *in vivo* elution test methods, and specifications to characterize the release rate of the drug, as well as some initial data to support the stability of both the drug substance and the polymeric carrier, if applicable. Additionally, adequate animal studies are needed to assess safety prior to going into human clinical trials for these products.

Common preclinical testing deficiencies

Some examples of the common deficiencies in applications received and reviewed by the FDA for the drug-coated stent include:

- Inadequate stent platform testing in terms of evaluating fatigue and corrosion testing. This testing is essential since it cannot be leveraged from previous experience with the bare stent platform alone. Adding the coating makes assessing the fatigue or cracking, and the effects of corrosion through potential cracks in the coating, extremely important.
- Inadequate analysis of surface modifications made to the device, either via application of the coating with the drug substance in it or. This relates to a lack of coating integrity, durability testing, and the characterization of drug content and its uniformity along the length of the stent producing a low-quality product.
- Incomplete *in vitro* pharmacokinetics in regards to the methods used for evaluating the drug elution from the stent. The FDA strongly recommends that the sponsors develop an *in vitro/in vivo* correlation whenever possible. This is extremely important in particular during the scale-up from a clinical trial batch or precommercialization batch to manufacturing the product for commerce.
- Additionally, it is also useful for the product to be well characterized, since if changes or improvements are made or need to be made to the product, the clearer the implication and evaluation of these changes in future clinicals will be.
- Chemistry and manufacturing control (CMC) issues are not often adequately addressed; stability and shelf life are very important. Device companies are used to a device paradigm where accelerated aging is compared to real-time aging to confirm these results have been accepted. In these cases, the Center for Device and Radiological Health (CDRH) has typically reviewed protocols, and when CDRH is comfortable with device materials and device packaging materials, these are straightforward protocols. The challenge occurs when the drug substance and a polymeric substance are introduced to the device; in this instance, many new issues arise that need to be considered in reference to the stability, including drug content, drug elution, impurities, and particulates of the combined product. It is recommended by the FDA that the sponsors follow the International Conference on Harmonization guidelines for evaluation of stability, especially for the drug substance in these instances.

Common preclinical animal study deficiencies

Some common preclinical animal study deficiencies are listed below using the drug-eluting stent as an example:

- Regarding animal studies submitted to the agency, the FDA often receives reports that do not provide adequate information to allow the agency to determine safety such that the risks and benefits are appropriate for the commencement of a clinical trial.
- In many instances, many applications lack information on adequate evaluation of the doses that were intended for use in the clinical study (since dose may change with stent size, the highest dose should be evaluated).
- The agency also requires evaluation of an overdosage to ensure that the FDA understands what the toxicity limits are in an animal model.
- Many reports do not include an evaluation of serial sections of myocardium. The FDA is interested in this, as well as arterial histopathology and necropsy reports for any deaths that might have occurred during the animal study.

From these it is clear that creating and executing a clear plan to characterize the actual product before clinical trial is crucial for combination product development.

Clinical investigation

For most combination products, one investigational application (investigational new drug [IND] or investigational device exemption [IDE] application) is submitted for the clinical investigation of the combination product as a whole. Generally, the regulatory guidance for INDs and IDEs provides substantial flexibility in considering how to address the issues posed by a particular product. Two such guidance documents that may be of interest to combination product developers are: (1) *Guidance for Industry, Investigators, and Reviewers Exploratory,* which provides an alternative for exploring candidate products during research and development prior to selecting the composition for further development, and (2) guidance on changes that may occur during investigational development of a device.

Clinical development questions are often asked regarding a trial design, sample size, the relevant number of clinical studies, statistical methods to be used, clear clinical endpoints, and appropriate indications or claims made on the final combination product. It is recommended that the technology behind the development of the combination product be looked at when making decisions on the sample size, which statistical techniques to use, endpoints measured and measurement methods for drug levels in areas not typically accessible, and techniques to evaluate drug-device interactions.

In regards to the above questions, the FDA encourages developers to seek early discussion with the agency around these concerns.

For combination products that include a device constituent part, it is necessary to evaluate the human factors of the device use on the safety and effectiveness of the product. Studies for the human factor effectiveness and safety would include evaluation of the use of the combination product by the user in a realistic, real-situation environment, which may also include the usual stressful conditions for the user. These studies would often include assessing all the components and accessories necessary to operate and properly maintain the device, for example, controls, displays, software, operation details, labeling, instructions for use, analysis of critical tasks, user error hazard, and risk analysis.

The human factors evaluations are recommended to take place early in the combination product development process so that the design features that need to be modified are identified earlier in the development process, prior to conducting any relevant studies that may determine the safety and effectiveness of the combination product.

Clinical evaluation

Using the drug-eluting stent as an example, the FDA looks for, primarily, a reasonable assurance of safety and effectiveness; it is critical to bear in mind that the clinical trial design should aim to meet both of these objectives. The usual standard of evidence for these products is a randomized controlled clinical trial. In terms of study endpoints for coronary drug-eluting stents, the FDA recommends that the primary endpoint or endpoints include at least one clinically meaningful endpoint.

The FDA would also evaluate the use of surrogate and co-primary endpoints. With approved drug-eluting stents now on the market from several different companies, there are many questions in regards to how a manufacturer could adequately design a study to be performed in the United States.

The coronary drug-eluting stent team, in cooperation with manufacturers and CDRH's Office of Biometrics and Surveillance, continues its efforts to develop reasonable clinical trial designs that will provide evidence of both effectiveness and, most importantly, safety. The use of independent core labs, clinical event committees, and an online, very active data safety monitoring board is critical and key to conducting a high-quality drug-eluting stent trial.

From a clinical research perspective, requirements for study design, sample size, and endpoint determination are not well defined. Generally, combination products involve more complex study designs and, as a result, longer clinical development timeframes. In some cases, the regulatory agencies may require larger trials for some products, which are not always aligned with market size or feasible patient enrollment rates. Additional delays may also occur as a result of unclear adverse event reporting standards or lack of respective guidance documents and strong endorsement of clinical holds in uncertain circumstances.

Clinical trials

Phase 1 clinical trials

The phase 1 clinical trial is conducted to determine safety and dosage using patients in the intended indication or healthy volunteers. The objectives of this trial are to define the maximal tolerated dose, understand side effects associated with the experimental drug, and assess the pharmacokinetic properties of the experimental drug. If possible, pharmacodynamic assessments are made to address the pharmacokinetic-pharmacodynamic relationship. Regulatory feedback on the trial design may be sought by requesting scientific advice at a pre-IND meeting. There are several categories of safety information that fall under the phase 1 IND requirements, including chemistry and manufacturing controls (CMCs). Both preclinical pharmacology/toxicology and systemic clinical exposure in normal subjects are required prior to beginning human investigations of the finished product. It is also important to note that if the drug substance has not been studied, this phase 1 IND safety information could very well influence the clinical trial design that would be necessary for the finished product. If there are toxicity issues or potential drug-drug interactions that are identified during the phase 1 safety information gathering, it may be necessary to alter the clinical study to look for those when evaluating the drug-eluting stent.

Example 1: A Device-Biologic Combination Product

- Before conducting trials, companies need to ensure applicability of the biologic with device (i.e., usability, potency, degradation, stability, etc.).
- All clinical evaluations of investigational devices, unless exempt, must have an approved IDE before the study is initiated.
- Device clinical studies consist of two phases (feasibility and pivotal) or three phases for drug. Feasibility studies are not always required for devices. Whether or not a feasibility study is required is based on the risk assessment of the device. Feasibility studies on the device side are equivalent to phases 1 and 2 for the pharmaceutical studies.
- Phase 1 studies of a combination product are sometimes not conducted in humans. Safety and dosage (operating range) can be determined using animal models and cadavers in some circumstances.
- When animal models are used in lieu of human clinical studies for device, companies need to validate these models (i.e., correlate animal models to human structure via scientific/structural relationship).

- Sterilization requirements need to be addressed where applicable prior to use in humans.
- Control trials are usually done in biologic clinical trials but not in device clinical trials.
- Companies need to consider the extent of development a device must have achieved prior to using it in humans. The regulatory strategy determines the requirement for a preclinical study to obtain an approved investigational device exemption (IDE) from the FDA prior to using a device in humans, or if the device will be used in clinical trials under the IND of an experimental drug. Biocompatibility must be demonstrated prior to using the device in humans.

Phase 2 clinical trials

The goal of phase 2 clinical trials is to assess the efficacy of the experimental drug in the population for which its use is intended, as well as to continue to examine its safety. Phase 2 is generally divided into phase 2a and phase 2b. In most instances, the decision to enter into full development is made in phase 2a, so the phase 2 trials need to be rigorously planned in order to support the decision for the development team to move on or terminate the program.

Example 2: A Device-Biologic Combination Product

- Phase 2 clinical trials in biologic development are for proof of concept and dose finding, whereas in device development there are no dosing-range studies.

Phase 3 clinical trials

The goal of phase 3 clinical trials is to demonstrate efficacy and expose a sufficient number of patients to the experimental drug for a sufficient period of time as defined in the ICH guidelines. These guidelines may be waived on a case-by-case basis following discussions with the appropriate regulatory agencies. In general, two well-controlled studies are required for regulatory approval of a given indication.

Example 3: A Device-Biologic Combination Product

- Certain markets require in-country testing of devices.
- A clinical readiness review must be conducted prior to pivotal studies.
- Studies may include labeling comprehension, human factors, validation of device specifications, and reading-level assessment.

- Clinical studies may be based on device equivalence if it is a 510(k) device.
- Pivotal and postmarket studies are not always required.
- Pivotal trials on the device are performed on production-equivalent units.
- Device clinical trials are documented in the design history file.
- Design verification on a device needs to be completed prior to phase 3 trials (pivotal study).
- There are no control studies in the clinical trials on a device.

Regulatory challenges

Combination products can present significant challenges in many companies not familiar with the regulatory requirement of a novel combination product. A combination product is especially challenging if it is the first one the company has developed and the staff's experience is exclusively with medical device regulations. The regulatory staff within a company developing a combination product needs to involve themselves as part of the team early on in the product development cycle. The regulatory affairs (RA) professional needs to get involved in the design process, evaluating the regulatory and clinical implications of the project as it evolves. The clinical and regulatory strategy planned by a company is critical to the development cycle and timely commercialization of the product for the indications sought.

The regulatory arena for combination products is evolving continuously, as agencies and industry obtain increasing knowledge about these products. The Office of Combination Products (OCP) is responsible for the prompt assignment of a new combination product to the lead FDA review center, which may be the Center for Devices and Radiological Health (CDRH), the Center for Drug Evaluation and Research (CDER), or the Center for Biologics Evaluation and Research (CBER). OCP has paved the way for companies to establish standardized and compliant development processes by recognizing the overlap between quality system regulations (QSR) and current good manufacturing practices (cGMP). OCP is additionally responsible for coordinating the premarket review process and the postmarket regulation of combination products.

How are combination products regulated?

A combination product is assigned to an agency center or alternative organizational division that will have primary jurisdiction for its premarket review and regulation. Under section 503(g)(1) of the act, assignment to a center with primary jurisdiction, or a lead center, is based on a determination of the primary mode of action (PMOA) of the combination product. For example, if the PMOA of a device-biological combination product is attributed to the biological product, the agency component responsible for premarket review of that biological product would have primary jurisdiction for the combination product. The lead center generally has responsibility for oversight of the regulation of the combination product, including the evaluation of current good manufacturing practice.

Section 503(g)(4)(D) of the act requires the FDA to "ensure the consistency and appropriateness of postmarket regulation of like [combination] products." To achieve consistency, the FDA will treat like combination products similarly. To ensure appropriateness, the FDA plans to require that manufacturers use the applicable current good manufacturing practice regulations for their combination products. In the regulation of a combination product, the application of consistent and appropriate current good manufacturing practice should help to ensure that the combination product is not adulterated under section 501 of the act, and is manufactured in accordance with appropriate regulatory provisions for the combination product and its constituent parts.

However, regulatory challenges can still be significant. For example, the FDA and industry are working to define the ground rules in areas such as product labeling and postmarket surveillance. Questions regarding product life cycle and postapproval changes still exist, including how to handle changes to the entire product versus components, changes across multiple companies, new indications, and unified versus separate labeling. In addition, it is still unclear how companies should manage differences across the various regulatory agencies for annual reporting, safety monitoring, and adverse event reporting. Companies have responded by increasing their involvement with the FDA. By consulting with the agency early and often, companies can determine the relevant marketing approval requirements (e.g., PMA or 510(k)s for devices, and new drug applications or biologic license applications for pharmaceuticals and biologics) in the product development process. Experts from the largest device and pharmaceutical companies are working closely with the agency to shape forthcoming guidance and requirements.

Figure 7.1 shows the interactions within the three FDA divisions for combination products. When combining products such as drugs or biologics and devices that are customarily developed using different regulatory paradigms, certain critical developmental issues, such as the interaction of the various components, in the regulation of combination products offer

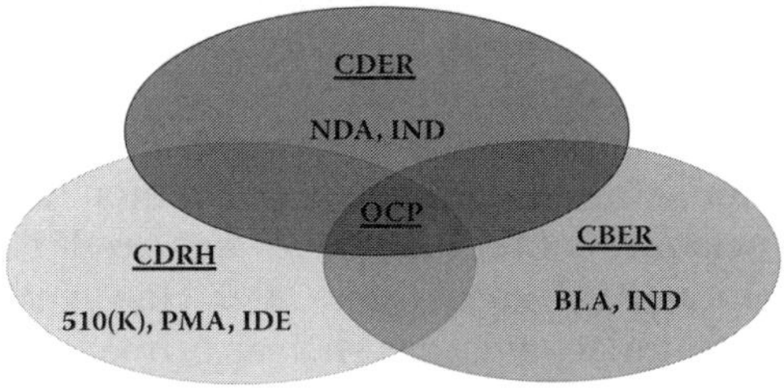

Figure 7.1 Interactions within the three FDA divisions for combination products.

challenges not typically seen with other types of medical products. Because each component product type is governed by a unique set of regulations, the differences in regulatory pathways for each component can impact the processes and documentation of all aspects of the product life cycle, including product development, manufacturing, and marketing. In addition, combination products increasingly use state-of-the art, innovative technologies that challenge existing regulatory and scientific knowledge. As the number and complexity of combination products increase, companies and regulatory agencies continue to struggle with their regulation.

The regulatory requirements resulting from product jurisdiction are still a challenge, as often it is unclear which GMP regulations are applicable during the manufacturing of a combination product and its inspection by the FDA. It is unclear how the assigned center will handle changes in manufacturing of the combination product due to differences in regulations governing the reporting of changes for biologics and devices.

For the development of biologic-device combination products, for example, therapeutic systems, such as bioartificial organs, it is difficult to determine the regulatory pathway for market approval of this product and clearer guidances are required from the FDA.

The HepatAssist system as a biologic-device combination product

As a combination product, the HepatAssist Liver Support System (LAS) could be designated as a device or a biologic. The importance of product designation and its implications is well demonstrated by the history of jurisdiction over LAS. There are several critical issues specific to LAS as a combination product: product processing, product storage and transportation, design and change control, product delivery to the patient, and various applicable regulations. Since LAS is a multicomponent system, the quality assurance/quality control (QA/QC) program was designed to cover device as well as biologic components. This program ensures the quality of cryopreserved cells, hollow-fiber cartridges, the hardware component, various disposables, and the clinical operations.

Uncertainty about how a product will be regulated and who must be dealt with creates increased risks and expense for the medical device company. Nevertheless, medical device manufacturers still must take an aggressive approach in establishing their regulatory strategy, or they may find their products or companies regulated out of business.

Preapproval issues

An incorrect regulatory strategy can cost not only money but time to market. Combination products present regulatory challenges. It is better to submit a Request for Designation (RFD) earlier than later. However, this should not be submitted until basic regulatory issues such as primary mechanism of clinical action and likely proposed intended uses have been resolved.

As we mentioned in Chapter 4 one of the first steps involved in the proposal of a combination product is to submit an RFD to the Office of Combination Products (OCP). The following information will usually be helpful to relay to the OCP: (1) a brief description of the product and its major components; (2) a description of the intended use of the product (the clinical/therapeutic claims, including the types of patients it will be used for); (3) an explanation of the product's modes of action (how the product works); and (4) any other information you think is relevant to your inquiry. The information outlined in 21 CFR 3.7 might also be appropriate. The assignment of the lead center is made based on a determination of the product's primary mode of action (PMOA). While the RFD is an important formal process, informal consultations can be very helpful. Companies should also have pre-RFD meetings with OCP for information and guidance. Tables 7.1 and 7.2, adapted from the FDA's August 2005 *How to Write a Request for Designation (RFD)*, provide examples of what to include in an RFD to the OCP, and also a checklist to ensure all the key information has been captured in the submission.

Premarket review

A combination product is assigned to an agency center that will have primary jurisdiction for its premarket review and regulation. Under section 503(g) of the act, the assignment of a lead center is based upon a determination of the PMOA of the combination product. For example, if the PMOA of a combination product is that of a biological product, then the combination product would be assigned to the agency component responsible for premarket review of that biological product, that is, CBER. Depending upon the type of combination product, approval, clearance, or licensure may be obtained through submission of a single marketing application, or through separate marketing applications for the individual constituent parts of the combination product. For most combination products, a single marketing application is sufficient for the product's approval, clearance, or licensure. In some cases, however, a sponsor may choose to submit two marketing applications for a combination product when one application would suffice. For example, a sponsor may choose to submit two applications in order to receive some benefit that accrues only from approval under a particular type of application (e.g., new drug product exclusivity, orphan status, or proprietary data protection when two firms are involved). In other cases, the FDA may determine that two marketing applications are necessary. For example, when one of the individual constituent parts of a combination product is already approved for another use, and where the labeling of the already approved product will need to be changed to reflect its new intended use in the combination product, the FDA may determine that two applications are necessary, if the labeling of the already approved product is subject to legal requirements different from those that will apply to the combination product. A guidance addressing the factors the FDA expects to consider in determining whether a single or multiple marketing applications should be

Table 7.1 Information to be Included in an RFD, per 21 CFR § 3.7(c)

21 CFR section	Information to be included
3.7(c)(1)	Sponsor identity, including company name, address, establishment registration number, company contact person, and telephone number
3.7(c)(2)	Product description, including: i. Classification, name of the product, and all component products, if applicable ii. Common, generic, or usual name of the product and all component products iii. Proprietary name of the product iv. Identification of any component of the product that has premarket approval, is marketed as not being subject to premarket approval, or has received an investigational exemption; sponsor identity; and the status of any discussions/agreements between the sponsors regarding the use of this product as a component of a new combination product v. Chemical, physical, or biological composition vi. Status and brief reports of developmental work, including animal testing results vii. Description of the manufacturing processes, including all component sources viii. Proposed use or indications ix. Description of all known modes of action, the sponsor's identification of the single mode of action providing the most important therapeutic action of the product, and the basis for that determination schedule and duration of use x. Dose and route of administration of drug or biologic xi. Description of related products, including their regulatory status xii. Any other relevant information
3.7(c)(3)	The sponsor's recommendation as to which agency component should have primary jurisdiction

Source: Adapted from Office of Combination Products, *How to Write a Request for Designation (RFD)*, Guidance for Industry and FDA Staff, 2005.

submitted for a combination product is in development and will be provided separately for public review and comment. Detailed information on the number of marketing applications required for combination products is discussed in Chapter 4.

In certain cases, it is not possible for either the FDA or the product sponsor to determine, at the time a request is submitted, which mode of action of a combination product provides the most important therapeutic action. Determining the PMOA of a combination product is also complicated when the product has two completely different modes of action, neither of which

Table 7.2 RFD Screening Checklist

Documents included in RFD package	Check
The identity of the sponsor, including company name and address, establishment registration number, company contact person, and telephone number (3.7(c)(1))	
Description of the product (3.7(c)(2))	
Classification, name of the product, and all component products, if applicable (3.7(c)(2)(i))	
Common, generic, or usual name of the product and all component products (3.7(c)(2)(ii))	
Proprietary name of the product (3.7(c)(2)(iii))	
Identification of any component of the product that already has received premarket approval, is marketed as not being subject to premarket approval, or has received an investigational exemption; the identity of the sponsors; and the status of any discussions or agreements between the sponsors regarding the use of this product as a component of a new combination product (3.7(c)(2)(iv))	
Chemical, physical, or biological composition (3.7(c)(2)(v))	
Status and brief reports of the results of developmental work, including animal testing (3.7(c)(2)(vi))	
Description of the manufacturing processes, including the sources of all components (3.7(c)(2)(vii))	
Proposed use or indications (3.7(c)(2)(viii))	
Description of all known modes of action, the sponsor's identification of the single mode of action that provides the most important therapeutic action of the product, and the basis for that determination (3.7(c)(2)(ix))	
Schedule and duration of use (3.7(c)(2)(x))	
Dose and route of administration of drug or biologic (3.7(c)(2)(xi))	
Description of related products, including the regulatory status of those related products (3.7(c)(2)(xii))	
The sponsor's recommendation as to which agency component should have primary jurisdiction based on the mode of action that provides the most important therapeutic action of the combination product (3.7(c)(3))	
For combination products where the mode of action that provides the most important therapeutic action cannot be determined with reasonable certainty, the sponsor's recommendation must be based on the assignment algorithm and an assessment of the assignment of other combination products the sponsor wishes the FDA to consider during the assignment of its combination product (3.7(c)(3))	
First step of assignment algorithm: assignment of other combination products presenting similar safety	
Second step of assignment algorithm: most expertise related to the most significant safety and effectiveness questions	

is subordinate to the other. To assign such products with as much consistency, predictability, and transparency as possible, the agency is issuing an algorithm to determine PMOA in those instances, to be codified at § 3.4(b). In those cases, the agency will assign the combination product to the agency component that regulates other combination products that present similar questions of safety and effectiveness with regard to the combination product as a whole. When there are no other combination products that present similar questions of safety and effectiveness with regard to the combination product as a whole (e.g., it is the first such combination product, or differences in its intended use, design, formulation, etc., present different safety and effectiveness questions), the agency would assign the combination product to the agency component with the most expertise to evaluate the most significant safety and effectiveness questions presented by the combination product.

When market approval authority is with CDER, drug approaches are necessary. Drug information required would be chemistry, manufacturing and controls, nonclinical pharmacology and toxicology, microbiology, and antimicrobials only. Sufficient information on the drug substance and the drug product include description, characterization, composition, manufacture, packaging, control, and specifications. Data on nonclinical pharmacology and toxicology in animal models are also necessary.

In order to complete a clinical study, an investigational new drug (IND) or an investigational device exemption (IDE) may be required. If it is an IDE, it could be reviewed by CDRH, or possibly CDER or CBER. One or two centers may review it. Once the study is completed, the marketing application may be submitted to CDRH as a premarket approval (PMA) or 510(k), or to CDER as an NDA, PMA, or 510(k), or to CBER as a biologics license application (BLA), PMA, or 510(k). The company may even be required to complete two separate submissions, or one "embedded" in another (e.g., a 510(k) embedded in an NDA).

User fees for combination products

What are user fees?

In 1992, Congress passed the Prescription Drug User Fee Act (PDUFA), PL 102-571. The user fee amendments were reauthorized by the Food and Drug Administration Modernization Act of 1997, and again by the Public Health Security and Bioterrorism Preparedness and Response Act of 2002. PDUFA authorized the FDA to collect fees from companies that produce certain human drug and biological products. When a company requests approval of a new drug or biological product prior to marketing, it must submit an application (e.g., new drug application [NDA] or biologics license application [BLA]) along with a fee to support the review process. In addition, companies pay annual fees for each prescription drug product marketed and for the establishment where the prescription drug product is manufactured.

The Medical Device User Fee and Modernization Act of 2002 (MDUFMA), PL 107-250, amended the act to provide for user fees for device reviews. The fees apply to certain premarket reviews of premarket approval applications (PMAs), product development protocols (PDPs), premarket reports (PMRs), biologics license applications (BLAs), and certain supplements.

How are application user fees determined for combination products?

As explained in the document *Assessing User Fees: PMA Supplement Definitions, Modular PMA Fees, BLA and Efficacy Supplement Definitions, Bundling Multiple Devices in a Single Application, and Fees for Combination Products; Guidance for Industry and FDA*, a combination product with a device component (e.g., a drug-device or biologic-device product) should be subject to the fee associated with the type of application required for the product's premarket approval, clearance, or licensure. For example, a biologic-device or a drug-device combination product for which a PMA is required should be subject to the PMA fee under MDUFMA, while a biologic-device or a drug-device combination product for which a 510(k) is required should be subject to the 510(k) fee under MDUFMA.

A biologic-device product regulated under section 351 of the PHS Act will be subject to the BLA fee under MDUFMA, if the biological component meets the definition of a device. Other biologic-device combination products (those with biologic components that do not meet the definition of a device), or drug-biologic combination products regulated under section 351 of the PHS Act, or drug-device or drug-biologic combination products regulated under section 505(b) of the act, that are human drug applications as defined in section 735 of the act, will be subject to prescription drug user fees. Prescription drug user fees may include application and yearly product and establishment fees.

Therefore, combination products for which a single marketing application is submitted should be subject to the fee associated with that type of application. Sponsors may be eligible for fee waivers or reductions (e.g., for small businesses) under PDUFA and MDUFMA. More information on available waiver options is provided below:

- In some circumstances, a sponsor may choose to submit two applications covering the various components of a combination product when one application would suffice. In such cases, two application fees would be assessed, one fee for each application. For example, a sponsor may choose to submit two applications when one would suffice in order to receive some benefit from having two applications (e.g., new drug product exclusivity, orphan status, or proprietary data protection when two firms are involved). Although sponsors may still be eligible for existing fee waivers or reductions in this circumstance, the sponsor

receives benefit by submitting two applications. Review of two applications when one would suffice places extra burden on FDA resources, and a user fee for each application would ordinarily be assessed.
- Likewise, when the FDA requires two applications for a combination product, two application fees would be assessed. Sponsors may be eligible for existing waivers or reductions under PDUFA or MDUFMA. In particular, the agency intends to look closely at whether a PDUFA "barrier to innovation" waiver may be appropriate to reduce the additional fee burden associated with the FDA's requirement for two marketing applications.

What user fee waivers are available under MDUFMA?
There are certain fee waivers available to manufacturers under MDUFMA:

- MDUFMA provides more limited user fee waiver options than PDUFA. Other than specific situations identified in Table 7.3 for which no application fee is required, standard MDUFMA fees are required for all device applications other than those from small businesses. Under MDUFMA, a small business is defined as one whose annual gross sales and revenue (for the firm and its affiliates) is ≤ $30 million.
- Under MDUFMA, small businesses pay 38% of the standard PMA and BLA fee and 80% of the standard 510(k) fee.
- MDUFMA also provides a one-time waiver for the first premarket application from a qualified small business.

Table 7.3 lists MDUFMA fee exemptions and waivers that can be utilized by manufacturers.

Combination product strategies

As we mentioned earlier, it is critical that companies start to approach these regulatory issues very early in the product development cycle. For example, if a company can assess if there are modifications or changes in direction that can be achieved at an early stage, it may save significant company resources. If it is determined that the regulatory requirements will be too high to justify the product's development, the resources can be moved to a project with more profit potential. The difference between getting to market through a 510(k) and an NDA can be hundreds of thousands of dollars and 10 to 15 years if the NDA route is taken. There are not many medical devices or manufacturers that can justify the latter approach.

The first step in this process is to determine if the product is a combination product. This starts with a review of the regulations and guidances discussed above. If it seems that they may apply, the next step is to research the competition. It is important to see how similar or competitive products were handled by the agency. This may give some indication as to what to

Table 7.3 MDUFMA Fee Exemptions and Waivers

Category	Exemption or waiver
Humanitarian device exemption (HDE)	Exempt from any fee
BLA for a product licensed for further manufacturing use only	Exempt from any fee
First premarket application (PMA, PDP, BLA, or premarket report) from a small business	One-time waiver of the fee that would otherwise apply
Third-party 510(k)	Exempt from any FDA fee; however, the third-party may charge a fee for its review.
Any application for a device intended solely for pediatric use	Exempt from any fee. If an applicant obtains an exemption under this provision, and later submits a supplement for adult use, that supplement is subject to the fee then in effect for an original premarket application.
Any application from a state or federal government entity	Exempt from any fee unless the device is to be distributed commercially

Source: Modified from FDA (OCP), *Application User Fees for Combination Products*, Guidance for Industry and FDA Staff, 2004.

expect, or a potential justification to change how the product is treated based on the precedents found. After the firm has a relatively good idea of how the product should be received, several agency review branches and centers can be informally contacted, or questioned through a third party.

If it becomes evident that the most likely regulatory path is inconsistent with the market potential of the product, the company may want to try to take actions that reduce the regulatory burden. The first action is to see if the labeling or claims can be adjusted, generalizing them so as not to be tied to a single drug. Another possibility is to investigate if there are other marketed drugs that are consistent with the labeled indication for the device. The packaging may be adjusted so as to separate the drug and device components. The indication may have to be reviewed and consideration given to changing the device indications to reduce the possibility that the product will conflict with a drug indication. There are design considerations that also may be appropriate and possible. For postmarketing considerations, segmenting drug and device manufacturing may help minimize some of the registration and GMP conflicts. The company may be forced to solicit a pharmaceutical or biologic company as a partner in order to jointly address the regulatory issues.

Once the design and labeling choices are set, the company should prepare to start negotiations with the agency. The FDA realizes that there are many types of products and different situations for which it has not developed guidance or a clear path. There may be precedents that it no longer considers current policy. It is in the company's best interest to prepare its strongest analysis and justification of the most reasonable and least burdensome regulatory path. Then the likely lead center should be formally approached in order to discuss whether there are any issues with the company's assessments. If not, a clear regulatory path to market has been identified, which should be confirmed in a presubmission meeting, as appropriate. If the FDA disagrees, a meeting still may be necessary in order to understand its position and most likely will involve other responsible centers.

At this point, if the differences cannot be resolved, it is time to involve the FDA ombudsman. An informal call may be warranted to gauge his/her understanding of the issues, and then a Request for Designation (RFD) needs to be prepared and submitted. Once the designation has been given, the company can proceed and comply with the decision. If the determined regulatory path is inconsistent with the business or market requirements of the product, the project should be dropped. Even if dropped, the exercise provides a valuable learning experience, giving the company excellent insight into the regulatory and policy positions of the agency. Regardless of the outcome, this is an extremely important role for a regulatory professional within a company. The regulatory requirements and accurate assessment of the regulatory direction for the combination product are a key part of the design and development process that must be managed by the regulatory professional. Effective regulatory strategy and the appropriate regulatory path can be as important to the success of the final product as any of the initial developement plans or performance specifications. We do realize that regulatory professional's decisions can be constantly second-guessed, especially if the outcome is not satisfactory to the company management.

Below are examples of regulatory filing strategies for some combination products:

1. *An implantable pump* to deliver an approved drug with the drug labeling providing information on dosing and delivery via an implantable pump. This pump was classified as a class III device and achieved marketing approval through a PMA. The PMOA was as a device to deliver the drug, not the action of the drug itself.
2. *A transdermal patch* with an approved drug to be delivered using a novel technology through the skin. Current drug labeling does not provide the delivery dose or mode. This product achieved its marketing approval as a drug by filing an NDA. In this instance, the patch technology had to investigate safety and efficacy of a novel method of drug delivery and dosage form.

3. *A heparin-coated cardiac catheter* was classified as a class II device and achieved marketing clearance via 510(k). The PMOA was as a device to open the blood vessel, not heparinize it.

In many cases companies prefer to have their combination products regulated as a device, even a class III with a PMA, since the resources during product development, clinical trials, and submission of the filings are often significantly lower than those required for an NDA. Usually a 510(k) or a PMA may take anywhere from 8 months to over 5 years; however, an NDA may typically require 3 to 15 years. Additionally, the challenges for filing an NDA need to include information on manufacturing, packaging, specifications, design of the stability program for the combination products, and, if novel product, development of specific test methods, determination of the lot size for the device component, biocompatibility issues between the drug and device, sterility, and so on.

The regulatory team member also needs to assess the progress of the regulatory strategy during development since although a single center may have primary jurisdiction over the combination product, the product could still fall into a category where interceptor consultations may be ongoing.

International regulations

With a combination product, the international regulatory hurdles coupled with the challenges inherent in combination products can very likely result in a greater challenge for the development team. Detailed regulatory requirements in various countries are discussed in Chapter 4.

The initial step in the determination of which European regulatory requirements would be applicable for a combination product is to determine the product designation. The correct product designation will drive the appropriate regulatory pathway. Straightforward classification can be easily achieved for traditional pharmaceutical drugs or medical devices; however, the classification of drug-device, drug-biologic, or biologic-device combination products requires thorough examination of the regulations of the nation or nations considered for marketing the product.

Following the initial classification of the product, it is extremely important to revisit and review the regulations against the product intended to be marketed. By excluding devices containing a human component, individual nations were forced to determine how such devices are regulated. In the United Kingdom, devices that contain a human component are easily missed and fall between the regulatory cracks, since these devices are considered neither a device nor a pharmaceutical; therefore, neither national regulatory agency has jurisdiction over the product. The Medical Devices Directive (MDD) has been amended by Directive 2000/70/EC of November 16, 2000, to include human components with medical devices. This amendment alters the exclusion referred to above and has the effect of allowing devices that incorporate human blood derivatives to be regulated by the MDD. According to the amendment, the notified body responsible for assessing the device will be required to seek a scientific opinion from the European Agency for the Evaluation of Medicinal Products (EMEA) on the quality and safety of the derivative, taking account of the appropriate community provisions and, in particular, by analogy, the provisions of Directives 75/31/EEC and 89/381/EEC. Under annex II, the notified body would have to give consideration to the opinion provided by the EMEA and could not deliver the certificate if the EMEA's scientific opinion was unfavorable. Additionally, under annex I the usefulness of the derivative as a part of the medical device would need to be verified taking account of the intended purpose of the device. Furthermore, under annex I, section 7.4, a sample from each batch of bulk or finished product of the human blood derivative must be tested by a state laboratory or a laboratory designated for that purpose by a member state. With each manufactured batch, the manufacturer must inform the notified body of the release of the batch of devices and send to it the official certificate concerning the release of the batch of human blood derivative used in the device, issued by a state laboratory or a laboratory designed for that purpose by a member state (annex II, article 1.4a).

In summary, a combination product brings many challenges to the regulatory professional responsible for obtaining marketing approval in Europe. With the many country-specific nuances and ever-changing environment, it is important to be aware of applicable requirements for products through current guidance documents and regulations accessible via the Internet, as well as through consultants or colleagues specializing in international regulatory affairs. Uncertainty about how a product will be regulated and who must be dealt with creates increased risks and expense for the medical device company. Nevertheless, combination product manufacturers still must take an aggressive approach in establishing their regulatory strategy, or they may find their products or companies regulated out of business.

Quality and compliance challenges

QSR versus GMP

Quality systems for devices and pharmaceuticals are different, but both are adequate within their scope. Once a combination product's regulatory pathway has been determined, that is, drug with a device component, or a device with a drug or biologic component, manufacturing and quality standards and processes need to be established. This can be challenging to quality and compliance professionals unless they proactively approach this in partnership with regulatory and product development professionals. Too often we find quality professionals focused on their traditional audit roles without realizing their potential in proactively enabling quality and compliance for combination products. They should help "design in" quality by applying quality engineering principles, developing and deploying quality and compliance strategies, and steering the product development team to achieve compliance.

A combination product classified as a drug needs to follow 21 CFR 210 and 211; a product classified as a device needs to follow the QSR found in 21 CFR 820. A realistic approach would be to have the device portion of the combination product adhere to the QSR standards ensuring applicable elements of the biologic or drug requirements are adequately addressed. Similarly, the biologic and drug component of the combination product needs to additionally address the QSR requirements, such as design control, corrective and preventive action program (CAPA), preventive maintenance, and so forth.

From a practical perspective, it is difficult and confusing to apply two conceptually similar but administratively different quality systems (e.g., device quality systems and pharmaceutical cGMP) within the same manufacturing facility. This should be avoided.

There are limited publicized good manufacturing practice regulations by the FDA currently for combination products. Until these regulations are well defined, it is our view that each constituent part (i.e., device, drug, or biological product) is subject only to its governing current good manufacturing practice regulations when marketed separately (see 21 CFR 3.2(e)(3) and (4)) and when manufactured separately as constituent parts of a combination that will later be combined (see 21 CFR 3.2(e)(1) and (2)). For example, if a drug is marketed that is intended for use only with an approved individually specified device that is also marketed separately, the drug constituent must comply only with 21 CFR Parts 210 and 211, and the device constituent must comply only with 21 CFR Part 820. For combination products produced as a single entity or copackaged (see 21 CFR 3.2(e)(1) and (2)), both sets of current good manufacturing practice regulations would be applicable during and after combining the constituent parts. We emphasize these both in this chapter as well as in other chapters simply because they are not only requirements for combination products, but also challenges that can be met only with disciplined execution.

Most frequently, companies would have their manufacturing facilities operate under one type of current good manufacturing practice system (i.e., that described by either the QS or cGMP regulation) based on their primary industry. Table 7.4 lists certain current good manufacturing practice provisions to consider during and after joining together copackaged and single-entity combination products. In many instances, the FDA recognizes and acknowledges that there is considerable overlap between the QS and cGMP regulations. As a result, it should not be necessary for manufacturers of combination products to maintain two separate manufacturing systems to ensure compliance with both sets of regulations during and after combining the constituents of the combination products together.

Table 7.4 Key Current Good Manufacturing Practice Provisions to Consider During and After Joining Together Copackaged and Single-Entity Combination Products

If the operating manufacturing control system is Part 820 (QS regulation)		If the operating manufacturing control system is Part 210/211 (cGMP regulation)	
Carefully consider these specific cGMP requirements	**Title**	**Carefully consider these specific QS requirements**	**Title**
§ 211.84	Testing and approval or rejection of components, drug product containers, and closures	§ 820.30	Design controls
§ 211.103	Calculation of yield	§ 820.50	Purchasing controls
§ 211.137	Expiration dating	§ 820.100	Corrective and preventative actions
§ 211.165		Testing and release for distribution	
§ 211.166		Stability testing	
§ 211.167		Special testing requirements	
§ 211.170		Reserve samples	

Source: *OCP, Current Good Manufacturing Practice for Combination Products, Draft guidance,* Guidance for Industry and FDA, 2004.

Note: Including all subsections, as appropriate.

During and after joining these types of combination products together, the FDA believes that compliance with both sets of regulations can generally be achieved by following one set because under a more general requirement in one set of regulations, it would be possible to develop and implement a practice that complies with a more specific requirement in the other set of regulations. To ensure consistent and appropriate current good manufacturing practice, it is recommended by the FDA that manufacturers of these types of combination products assess how best to comply with both sets of regulations, during and after joining the constituent parts together, by carefully considering the requirements of the cGMP and QS regulations in relation to the constituent parts, and the combination products they manufacture.

Additionally, for a biologic-device or a drug-device combination product, other specific requirements to ensure compliance with both the cGMP and QS regulations may need to be considered. An example can be including aseptic control assurance for drug and biological product constituent parts unable to withstand terminal sterilization (21 CFR 211.113(b) and § 211.42).

Before combination or copackaging, the manufacture of each constituent part is subject only to the current good manufacturing practice regulations associated with each constituent part. For example,

- For a drug-coated device, the drug constituent for a drug substance or active pharmaceutical ingredient part would be subject only to the cGMP regulation (or to section 501(a)(2)(B) of the act for a bulk in pharmaceutical-biologics); microbiological tests are typically validated and executed by the sites. Analytical methods developed in R&D must be eventually developed into commercial test methods. The device constituent part would be subject only to the QS regulation. However, after combination of the constituents, both sets of regulations would apply to the combination.
- For a photodynamic therapy system consisting of a laser and a photosensitizing drug that are marketed separately, the laser would be subject to the QS regulation while the photosensitizing drug would be subject to the cGMP regulation.

Drug-eluting cardiovascular stent

Using the drug-eluting cardiovascular stent (DES) as an example, how does a primary device company address the needs of the company's current quality system to comply with U.S. FDA requirements for the combination product?

There are key fundamental differences between the pharmaceutical GMP regulations (21 CFR 210 and 211) and the device quality system regulation (QSR) in 21 CFR 820.

The differences in the physical properties for drug components versus medical device components have resulted in at least two different ways to control the manufacture of these product types.

Drug versus device components

Drug components are generally less stable than device components, and hence may be prone to degradation resulting from humidity, heat, contamination, and microbial or other chemical action.

In contrast, medical device components such as metal or plastic parts are usually inert and may be susceptible to magnetic interference, and so on. The drug GMP regulations have a more stringent requirement for the control and monitoring of environmental conditions, microbial or chemical contamination, and material handling in comparison to the device QSR.

Additionally, labile drug components, in-process materials, or finished products usually require an expiration date to ensure the components' quality and stability. The expiration dates for the drug components are usually provided by the supplier; however, expiration dates for the in-process materials and finished products need to be determined by the manufacturer of the combination product. Hence, stability testing needs to be performed by the company via use of analytical laboratory testing. Since this is not a typical operation for a device company, the company needs to factor this stability requirement into the timeline for introduction of the combination product.

The device company also needs to factor in additional verification by personnel during formulation operations since many drug components are white powders, which could result in mix-ups during operations. There are no double-check requirements in the device QSR.

The device company, as a result, may need to modify or build a manufacturing operation site to comply with the pharmaceutical GMP requirements of its drug component for the combination product. Medical device firms have not needed to perform extensive chemical analysis for incoming components, and having to comply with the drug portion of the combination product, this is one of the costliest operations. The stability program would also demand significant resources for recruiting skilled staff, developing reliable stability assays, and so on.

An additional challenge for the device manufacturer would be to define areas for quarantining goods such as incoming components, in-process materials, or finished goods. The facility water also needs to be tested for portability according to the Environmental Protection Agency (EPA) Primary Drinking Water Standards. Also, temperature-controlled storage areas for the drug component need to be addressed.

Application of cGMP regulations to combination products

In general, what determines which cGMP regulations apply to a combination product? Per the FDA's proposed guidance of January 15, 2007, it is suggested that first and most basic, each constituent part of a combination product is subject to its governing cGMP regulations before the constituent parts are combined, merged, or joined. After the constituent parts are combined, merged, or joined, the cGMP regulations under which the manufacturer currently operates are the regulations that principally would govern the combination product.

The manufacturer should also assess whether its existing quality system is adequate to ensure the quality of the product. In determining whether a current system is adequate, manufacturers should evaluate various factors, such as whether the manufacturer already has systems in place that address a particular requirement of another set of cGMP regulations.

The guidance also suggests, in instances where certain facilities may not have a cGMP system in place (e.g., a start-up company or new facilities of an established manufacturer), that the company implement the cGMP regulations that correspond to the PMOA of the combination product that will be manufactured at the facility. For example, if the PMOA is that of a drug, the facility should implement Parts 210 and 211.

Additionally, the manufacturer of the finished combination product has a responsibility to ensure the overall quality of the product. As part of this obligation, the manufacturer should ensure that the cGMP compliance of third-party service providers that perform manufacturing functions is adequate.

In general, the GMP strategy for a company in process of defining the quality system for its relevant combination product would need to consider several areas. The organization needs to develop a GMP/quality system plan suitable for its combination product, addressing the following key elements:

- Calculation of yields
- Testing and approval or rejection of components
- Drug product containers and closures
- Expiration dating
- Testing and release for distribution
- Stability testing
- Special testing requirements
- Reserve samples

The biological product regulations, 21 CFR Parts 600–680, may also apply to the manufacture of drugs that are biological products, along with the drug cGMP provisions. There is significant overlap in the cGMP and QS regulations, for example, both regulations establish requirements for the following elements:

- Management, organization, and personnel
- Documentation and record keeping
- Flexibility in application to the manufacture of particulate products

However, each set of regulations is somewhat different, as each is somewhat tailored to the characteristics of the types of products for which they were designed for by FDA; this can be seen in Table 7.4, for example, cGMP for drugs or biological products, QS regulation for devices. Each set of regulations also contains certain specific requirements that may be only more generally described in the other regulation. Usually, these specific requirements are related to the unique characteristics of a drug, device, or biological product. For example:

1. *Calculation of the yield and stability for a drug constituent part* (21 CFR 211.103): The cGMP regulation has specific requirements for the calculation of yield and for ensuring stability of the drug product; however, for a combination product with a drug constituent part, the QS regulation yield and stability requirements would be incorporated more generally under 21 CFR 820.30(g) as part of the design validation provisions.
2. *Corrective and preventive action (CAPA)* (21 CFR 820.100): The QS regulation has detailed CAPA requirements, while CAPA principles are more generally identified in the cGMP regulation as part of production record review (21 CFR 211.192).

Table 7.5 lists the differences in device, drug, and biologic regulations to be considered for a combination product.

Pharmaceutical companies need to consider the following key GMP provisions during and after joining together copackaged and single-entity combination products:

- Design control procedures must be developed and followed, which includes design reviews at designated points throughout development.
- Design control activities begin after the feasibility stage of development.
- Purchasing controls procedures need to be developed and implemented.
- A corrective and preventive action program (CAPA) needs to be developed and implemented. The CAPA system must include a method for determining the effectiveness of corrective actions implemented.
- Management reviews need to take place. This review can assess the effectiveness of the quality system.

Per the FDA's guidance on hybrid cGMP systems, determining whether an existing cGMP system is adequate is more challenging when a facility joins, merges, or combines two or more constituent parts into a single entity or kit combination product. A facility's existing cGMP system will be the overarching cGMP system in this situation. That said, a manufacturer should also assess whether the existing system at a facility is adequate to ensure the quality of the finished combination product. Quite often a facility that manufactures single-entity or kit combination products will need to implement provisions of cGMP regulations other than those that are already implemented at the facility in order to have an adequate system.

Furthermore, due to the differences among cGMP regulations and the differences in the types of combination products, the specific analysis

Table 7.5 Table Representing Regulations to be Considered for a Device, Drug, or Biologic Combination Product

Device QSR(21 CFR 820)	Drug cGMP(21 CFR 210 and 211)	Biologics (CFR 600, 601, 610, etc.)
a. General provisions (scope/definitions). Bio(CFR 600) (definitions).	a. General provisions (scope, definitions, quality system)	a. General provisions (scope/definitions)
b. Organization and personnel (responsibilities of QC unit, personnel qualifications, personnel responsibilities, consultants). Establishment standards (physical establishments, equipment, animals, records, retention samples, reporting of biological product licenses' manufacturers, temperatures during shipping).	b. Quality system requirements (management responsibility, quality audit, personnel including consultants)	b. Quality system requirements (management responsibility, quality audit, personnel including consultants)
c. Buildings and facilities (design, construction, lighting, ventilation, air filtration, HVAC, plumbing, sewage and refuse, washing and toilet facilities, sanitation, maintenance). Bioestablishment inspection (inspectors, time of inspection, duties of inspector).	c. Design controls	c. Design controls
d. Equipment (design, size, location, construction, cleaning and maintenance, automatic, mechanical, and electronic equipment, filters). Bioreporting of adverse experiences (postmarketing reporting of adverse experiences, distribution reports, waivers).	d. Document controls	d. Document controls
e. Control of components and drug product containers and closures (receipt and storage, testing and approval/rejection, use of approved components, containers, and closures, retesting, rejected components, containers, and closures).	e. Purchasing controls	e. Purchasing controls

(Continued)

Table 7.5 (*Continued*)

Device QSR(21 CFR 820)	Drug cGMP(21 CFR 210 and 211)	Biologics (CFR 600, 601, 610, etc.)
f. Production and process controls (written procedures, deviations, change in components, yield calculation, equipment identification, sampling and testing of in-process materials and products, time limitations on production, control of microbiological contamination, reprocessing)(610-f). Dating period releases (date of manufacture, dating periods for licensed biological products).	f. Identification and traceability	f. Identification and traceability
g. Packaging and labeling control (materials examination and usage criteria, labeling issuance, packaging and labeling operations, tamper-resistant packaging, drug product inspection, expiration dating). (610-g) Labeling standards (container label, package label, legible type, divided manufacturing to be shown, name and address of distributor, products for export, bar code label requirements).	g. Production and process controls (production and process controls, inspection, measuring, and test equipment, process validation)	g. Production and process controls (production and process controls, inspection, measuring, and test equipment, process validation)
h. Holding and distribution (warehousing procedures, distribution procedures).	h. Acceptance activities (receiving, in-process, and finished device acceptance, acceptance status)	h. Acceptance activities (receiving, in-process, and finished device acceptance, acceptance status)
i. Laboratory controls (general requirements, testing and release for distribution, stability testing, special testing requirements, reserve samples, laboratory animals, penicillin contamination).	i. Nonconforming product	i. Nonconforming product

(*Continued*)

Table 7.5 *(Continued)*

Device QSR(21 CFR 820)	Drug cGMP(21 CFR 210 and 211)	Biologics (CFR 600, 601, 610, etc.)
j. Records and reports (general requirements, equipment cleaning and use log, component and drug product container, closure and labeling records, master production and control records, production record review, laboratory records, distribution records, complaint files).	j. Corrective and preventative action	j. Corrective and preventative action
k. Returned and salvaged drug products.	k. Labeling and packaging control (device labeling, device packaging)	k. Labeling and packaging control (device labeling, device packaging)
	l. Handling, storage, distribution, and installation	l. Handling, storage, distribution, and installation
	m. Records (general requirements, device master record, device history record, quality system record, complaint files)	m. Records (general requirements, device master record, device history record, quality system record, complaint files)
	n. Servicing	
	o. Statistical techniques	o. Statistical techniques

undertaken by a manufacturer is different, depending upon the facility's existing cGMP system and what type of combination product is manufactured. Each set of cGMP regulations contains key elements based upon the unique characteristics of the types of products the regulations were designed to address. Table 7.6 sets forth these unique elements.

Facility, infrastructure, and manufacturing challenges

In Chapter 5 we discussed the requirements for manufacturing combination products. In the manufacturing of a combination product, scale-up and quality management are important considerations during development. The

Table 7.6 GMP Regulations That are Different Across Device/Drug/Biologic Systems

Devices	Drugs and most biological products	Certain other biological products
Design controls (820.30) Purchasing controls (820.50) CAPA (820.100)	Containers and closures (211.84) Calculation of yield (211.103) Aseptic control assurance for constituent parts unable to withstand terminal sterilization (211.113(b) and 211.42) OTC drug constituent parts (211.132) Expiration dating (211.137) Testing and release for distribution (211.165) Stability testing (211.166) Special testing requirements (211.167) Reserve samples (211.170)	Additional requirements as they apply under 21 CFR Parts 600–680

manufacturing methods and processes affect both premarket development and postmarket regulation of the combination product.

As a result, manufacturers of combination products have to especially pay attention to the effect of the manufacturing methods on the interaction of the constituent parts. In an example of a device-biologic combination product:

- *The stability* of the final combination product as a whole may be significantly different from the original, separate constituent parts.
- *Sterilization* may also have a significant impact on a drug or biological product constituent parts that may be altered or destroyed by terminal sterilization techniques. The manufacturer of the finished combination product needs to ensure that the product has reached a sterile state, and also all materials used in the packaging, and so on, are compatible with the sterilization process used.
- For constituent parts that use *aseptic manufacturing techniques*, developers are encouraged to implement manufacturing techniques to ensure aseptic control for the combination product.

Additionally, once the preclinical and clinical studies begin, any changes to the manufacturing process for the drug, biologic, or device constituents, or for the combination product, may affect the safety or effectiveness of the finished combination product. For example, changes in concentration, inactive ingredients, software, or the methods of the constituent parts could affect the performance characteristics of the combination product.

Additionally, the applicable device constituent design controls would consider anticipated manufacturing changes during investigational development. In order to address such manufacturing considerations, it may be necessary to develop new manufacturing techniques, in-process testing, testing specifications, and other characterization methods to assess changes in the constituent parts and for the combination product as a whole. For certain developmental changes, additional bridging studies (*in vitro*, preclinical, or clinical) may be appropriate.

In addition to considering manufacturing changes that may occur during premarket development, the FDA also recommends early consideration of anticipated postmarket manufacturing changes for the combination product or its constituent parts. The FDA encourages manufacturers to establish arrangements with the manufacturers of constituent parts to maintain sufficient awareness of manufacturing changes in constituent parts that may occur during the premarket or postmarket period. Such awareness could help to ensure continued safety and effectiveness of the combination product by ensuring that the potential impact of a manufacturing change is evaluated in a manner appropriate for the stage of combination product development. As appropriate, these postmarket manufacturing changes may require careful review, validation, and prior approval before marketing. For some products, it may be helpful to develop postapproval change protocols for further discussion with the agency.

For example, for a biologic-device combination product, some of the challenges can be product processing, product storage and transportation, design and change control, product delivery to the patient, and various applicable regulations. Since a biologic-device combination product is a multicomponent system, the quality program needs to cover device as well as biologic components. Examples of manufacturing controls to be used for such a product would include the following regulations:

1. 21 CFR Part 210: cGMP in Manufacturing, Processing, Packaging, or Holding of Drugs
2. 21 CFR Part 211: cGMP for Finished Pharmaceuticals
3. 21 CFR Part 600 and 610: General Biological Products Standards
4. 21 CFR Part 820: Medical Devices

Postlaunch challenges

Once a combination product is designed, developed, and launched successfully, it is important for the manufacturer to maintain a disciplined approach to ensure that the combination product quality and reliability is not impacted negatively due to variations in incoming material, manufacturing environment, and so on. Process control and monitoring is important on an ongoing basis. Maintaining adequate documentation can be challenging but is extremely important to ensure traceability, and so on.

Often, the manufacturing costs for combination products at launch can be higher since companies would prefer to be first to the market. However, in order to reduce these costs, companies might resort to implementing Lean manufacturing techniques, transferring to a low-cost location, and so forth. Caution must be exercised to ensure that business-value-added activities such as quality inspection are not minimized as a result of these efforts. In addition, companies must take into consideration the total cost of the product throughout its life (including costs of revalidation, resubmission, etc.).

In this chapter, we provided the reader with an extensive look at various challenges posed in developing, manufacturing, and selling combination products. We encourage the reader to pay attention to other challenges that are not mentioned in this chapter. We strongly believe that as with any new endeavor, dealing with combination product–related challenges will become easier with sustained leadership, expertise, caution, resources, and so forth.

Bibliography

21 CFR, Part 210. *cGMP in manufacturing, processing, packaging or holding of drugs*, 2006.

21 CFR, Part 211. *cGMP for finished pharmaceuticals*, 2006.

21 CFR, Part 600. General biological products standards, 2006.

21 CFR, Part 610. General biological products standards, 2006.

21 CFR, Part 820. *Quality system regulations for medical devices*, 2006.

Application user fees for combination products. Guidance for industry and FDA staff, FDA, April 2005.

Boam, Ashley B. Drug-eluting stents: an FDA case study. *Regulatory Affairs Focus*, pp. 18–23, November 2003.

Broderick, Julie N. Brave new world: EU combination products regulation. *Regulatory Affairs Focus*, September 2006.

Current good manufacturing practice for combination products. Draft guidance for industry, FDA, 2004. www.fda.gov/oc/combination/OCLove1dft.html

Current good manufacturing practice regulations for combination products. *Federal Register* 22565, April 24, 2006.

Diamond, Mason W. Beyond primary mode of action: a new look at "chemical action" and its impact on combination product designations. *Regulatory Affairs Focus*, September 2006.

Early development considerations for innovative combination products. Guidance for industry and FDA staff, Office of Combination Products, September 2006.

Final rule on the definition of primary mode of action of a combination product. *Federal Register* 49848, August 25, 2005.

Garrison, Connie. Global biologics, combination products in Europe: a case study. *Regulatory Affairs Focus*, March 2002.

Guidance to industry: changes or modifications during the conduct of a clinical investigation, final guidance for industry and CDRH staff. FDA, CDRH, May 29, 2001.

How to write a request for designation (RFD). Guidance for industry and FDA staff, FDA, August 2005.

IND meetings for human drugs and biologics chemistry, manufacturing, and controls information. Guidance document, FDA, May 2001.

ISO 10993-1: Biological Evaluation of Medical Devices—Part 1: Evaluation and Testing, 2nd ed. International Organization for Standardization, 1997.
Kramer, Mark D. Combination products: challenges and progress. *Regulatory Affairs Focus,* pp. 30–35, August, 2005.
Patel, Subhash. Combination products: puzzles of regulations. *Regulatory Affairs Focus,* September 2006.
Pitkin, Zorina. Regulation of biologic/device combination products. *Regulatory Affairs Focus,* June 2000.
Q1D bracketing and matrixing designs for stability testing of new drug substances and products. FDA, CDER, 2003.
Sall, Barry S., Lassoff,, Peter and Babbitt, Bruce. Getting started with a combination product: Part I, Overcoming the regulatory challenges in this ever-expanding field yields rewards. *Medical Device & Diagnostic Industry,* March 2003.
Submission and resolution of formal disputes regarding the timeliness of premarket review of a combination product. FDA: www.fda.gov
Swain, Eric. Forging new regulatory pathways at FDA. *Medical Device & Diagnostic Industry,* September 2004.
Vinhais, Joseph. Integrating compliance-based solutions for the enterprise. *Regulatory Affairs Focus,* September 2006.

chapter eight

Postlaunch compliance requirements

In Chapter 7, we discussed various challenges that can be faced and pitfalls to be avoided by companies developing and manufacturing combination products. We touched upon postlaunch and compliance challenges. In this chapter we will focus on postlaunch compliance requirements, including, but not limited to, requirements in the areas of postmarket surveillance, tracking and reporting, and postmarket changes. Certain concerns facing companies are indicated below:

- How will the facility be registered?
- Will the company be required to comply with the drug good manufacturing practices (GMP), the device quality system regulations (QSR), or both?
- What type of vigilance reporting is required? adverse drug reports (ADRs), medical device reports (MDRs), or both?
- If a change is made to the device, what rules determine whether the Food and Drug Administration (FDA) must be notified?
- Does the change need to be approved before implementation?
- What rules need to be followed on traceability, device tracking, and so forth?
- How is advertising and promotion going to be regulated?

The postmarket safety reporting requirements for drugs, devices, and biological products share many similarities. For example, each requires reports of deaths and serious adverse events, as well as submission of periodic or follow-up reports. The FDA believes that for most combination products, appropriate postmarket safety reporting may be achieved by following the regulatory provisions associated with the type of marketing application used for their approval/clearance. Nonetheless, the reporting requirements for drugs, devices, and biological products each have certain unique requirements based upon the products for which they were designed. An example can be observed in a drug-device combination product approved or cleared under the device provisions and ordinarily subject to medical device reporting (MDR) under 21 CFR Part 803. If the product was subject solely to Part 803, however, some of the reporting requirements that would have been applicable to the drug constituent part of the combination product, had that drug been regulated under the drug provisions of the act (and subject to postmarket reporting under 21

CFR 314.80), would not ordinarily apply. Conversely, had the same combination product been regulated under the drug provisions of the act, some of the reporting requirements that would have been applicable to the device component of the combination product would not ordinarily apply.

In many instances, the FDA and industry are working together to define the ground rules and requirements in areas such as product labeling and postmarket surveillance. In many cases, there are significant questions that revolve around product life cycle and postapproval changes. For combination products, these include how to handle changes to the entire product versus individual components, changes across multiple companies, new indications for use, unified versus separate labeling efforts, who files the adverse event reports and in what timeframes, who submits the complaint reports, and to which branch of the agency. Additionally, for companies involved in combination product development and launch, it is still unclear how they should manage differences across the various regulatory agencies for annual reporting, safety monitoring, and adverse event reporting for their products.

As we mentioned in the previous chapter, in many instances, companies have significantly increased their involvement with the FDA in heavy discussions by means of consulting with the agency early and frequently. As a result, these companies can often determine the relevant marketing approval requirements (e.g., premarket approval application [PMA] or 510(k) for devices, and new drug applications or biologic license applications for pharmaceuticals and biologics) in the product development process. Experts from the large device and pharmaceutical companies are working closely with the agency to help shape future guidance and requirements.

The postlaunch compliance requirements for drugs, devices, and biological products have many similarities, which entail review and monitoring of customer complaints, tracking and reporting adverse events, conducting overall postmarket surveillance on the product to ensure safety and efficacy, incorporating customer feedback on the design and performance of the product to improve upon the next generation, and so on.

FDA's role in regulation of products

The FDA has established three centers: Center for Drug Evaluation and Research (CDER), Center for Biologics Evaluation and Research (CBER), and Center for Devices and Radiological Health (CDRH). Among these, the 1938 federal *Food, Drug, and Cosmetic Act* (*FD&C Act*) is the common foundation, as drugs, devices, and biologics share many basic features. However, drugs, devices, and biological products also have certain unique and specific reporting requirements for premarket application, labeling, GMP regulations, postmarket safety reporting, design controls, and other requirements.

These differences and specific postmarket requirements present an even greater challenge in the regulation of combination products. Additionally, diversity of combination products and varied, advanced technologies make matters even more complex. As a result of emerging and complex product designs for combination products with increased indications for use, the need for the FDA Office of Combination Products (OCP) to assign regulatory jurisdiction to a lead center has become even more critical. Accurate designation of the constituent comprising the primary mode of action (PMOA) by the OCP is critical for these companies, since the time, complexity, developmental costs, and risks associated with the novel product are usually tied to the regulatory pathway. The PMOA is defined as the single mode of action of a combination product that provides the most important therapeutic action. To make that determination, both industry and the FDA must rely on, and to some extent be constrained by, the definitions of a drug, device, and biologic as stated in the *FD&C Act*.

Section 201(h) of the *FD&C Act* defines a device as

> an instrument, apparatus, implement, machine, contrivance, implant, *in vitro* reagent, or other similar or related article, including any component, part, or accessory, which is intended to affect the structure or any function of the body of man or other animals, and which does not achieve its primary intended purposes through chemical action within or on the body of man or other animals and which is not dependent upon being metabolized for the achievement of its primary intended purposes.

As stated, the law implies that the term *chemical action* defines a biologic or drug effect and impacts not only the development of combinations under the current system, but assignments of combination products to the various centers, which are made by the OCP via a Request for Designation. Determinations of the device attributes of a noncombination product are made via a 513(g) submission to the Center for Devices and Radiological products, but novel devices as well.

Medical devices are subject to medical device reporting (MDR) under 21 CFR Part 803. Drug products are subject to postmarket reporting under 21 CFR 314.80, and biological products are subject to the Vaccine Adverse Event Reporting System (VAERS) under 21 CFR 600.80(c)(1). Listed below are examples of the FDA's safety reporting requirements for postmarket for medical devices, drugs, and biologics, respectively.

Medical devices

Postmarket monitoring

Companies need to actively follow their device performance in the market. The information obtained from postmarket data would complement information and data obtained during premarket clinical studies. The challenges faced by companies upon commercial distribution of devices may be different than those faced at clinical trials. Upon launch of a device, companies are faced with issues such as long-term safety outcomes, patient populations with greater disease diversification, intended or accidental misuse of a device, and unexpected device malfunctions. Postmarket data on a product can therefore be extremely valuable for companies to implement appropriate corrective action on the design or manufacture of the original device. These data can be obtained through systematic efforts to monitor the product after launch. Postmarket monitoring should include sales and marketing input, service reports, quality test results, internal audits, clinical investigations, postmarket surveillance studies, medical literature, and regulatory/enforcement activities. Companies also rely on the complaint handling process to detect current or potential product problems. Typically, companies take a risk management approach on a released product—beyond the initial risk assessment during the development phases—to evaluate how modifications to the current design can enhance the second-generation product and make suitable changes.

A postmarket surveillance system is a means for a company to receive warnings and signals regarding the performance of its product in the field for early detection of potential problems. The FDA has authority over the following activities over companies:

- Require postmarket clinical studies as a condition of product approval for class II and III devices.
- The 1990 Safe Medical Device Act required postmarket surveillance studies for all devices marketed after January 1, 1991, that met the criteria of:
  - Permanent implants with serious health consequences or death if they failed
  - Devices intended for use in supporting or sustaining human life
  - Devices whose failure presented a potentially serious risk to human health

The Center for Devices and Radiological Health (CDRH) reported in March 2005 that only 58% of clinical condition-of-approval studies required for PMAs under 21 CFR 814.82, granted between January 1, 1998, and December 31, 2000, were completed or progressing on time, and announced plans to implement changes to improve postsurveillance tracking and compliance. Additionally, recent legal issues, arising in the drug industry (e.g., for Vioxx), most likely will also further impact device and drug companies alike, by heightening postmarket surveillance requirements by companies. Even if postmarket clinical studies are not ordered, medical device companies need to consider including an additional postrelease phase as part of their design control programs.

Postmarket adverse event reporting

The key differences in the drug, device, and biological product postmarket safety reporting regulations that the FDA currently believes are the most significant to monitor and assess are as follows:

1. Device malfunction reporting (21 CFR 803.3(r)(2)(ii), 21 CFR 803.20):
    - In addition to the reporting of device malfunctions associated with a death or serious injury, the MDR regulation also requires reporting of device malfunctions where no death or serious injury has occurred, but when such device or similar device marketed by the manufacturer would be likely to cause or contribute to a death or serious injury, if the malfunction were to recur.
    - Reporting for drugs and biological products *does not* include the analogous requirement, that is, reporting of product failure that *could* result in a death or serious injury, but for which a patient event did not occur.
    - In order to ensure consistent and appropriate postmarket regulation for some combination products with device constituent parts regulated under the drug or biological product provisions of the act, device malfunction reporting may be necessary.
2. Five-day MDR reporting (21 CFR 803.10(c)(2)(i)): The MDR regulation in this instance requires reporting of the following:
    - Any reportable event that necessitates remedial action to prevent an unreasonable risk of substantial harm to the public health.
    - Any MDR reportable event for which the FDA has made a written request for the submission of a 5-day report.
    - To ensure consistent and appropriate postmarket regulation, for some combination products regulated under the drug or biological product provisions of the act, 5-day MDR reports may be necessary.
3. Drug and biological product "alert" reporting (21 CFR 314.80(c)(1) and 600.80(c)(1)): For drugs and most biological products, postmarket safety reporting emphasizes:
    - Adverse events that are both serious and unexpected. Although device safety reporting requires 30-day reports of any serious injury, the reports would not necessarily flag an event as both serious and unexpected, and they would be submitted at 30 days rather than the earlier alert reporting period of 15 days.
    - In order to ensure consistent and appropriate postmarket regulation, for some combination products with drug or biological product constituents that are regulated under the device provisions of the act, such alert reporting may be necessary.

- Postmarketing 15-day alert reports. The applicant shall report each adverse drug experience that is both serious and unexpected, whether foreign or domestic, as soon as possible but no later than 15 calendar days of initial receipt of the information by the recipient.
- Postmarket 15-day alert reports follow-up. The applicant shall promptly investigate all adverse drug experiences that are the subject of this postmarketing. The applicant shall submit follow-up reports within 15 calendar days of receipt of new information or as requested by the FDA.

4. Blood-related deaths (21 CFR 606.170): The biological product regulations require reports to be submitted to CBER as soon as possible (e.g., by phone, fax, or e-mail), with a 7-day written report of any blood-related death. The FDA believes that for some blood-containing combination products regulated under the device or drug provisions of the act, such early notification of blood-related deaths may be necessary in order to ensure consistent and appropriate postmarket regulation.

Postmarket safety reporting for combination products

Due to the unique and specific reporting requirements for drugs, devices, and biological products, combination product postmarketing safety reporting requirements depend upon the provisions of the act under which the product was cleared or approved. Which center holds primary jurisdiction for the product, and how many applications were submitted?

Consistent and appropriate combination product postmarket regulation and appropriate ongoing assessment of risks associated with a combination product are areas addressed by the FDA. The agency currently is considering mechanisms by which the postmarket safety reporting requirements ordinarily associated with the product's marketing application may be supplemented, as appropriate, to take into account the product's combined components.

One idea is a reporting scheme under which the same postmarket safety reports would be submitted for any combination product, regardless of the marketing application used for its approval or clearance. However, the FDA believes that the current drug, device, and biological postmarket safety reports highlight the differences in the product types, identified below, while monitoring and assessing the risks associated with combination products.

To ensure consistent and appropriate postmarket regulation of combination products, and to ensure an appropriate ongoing assessment of the risks associated with a combination product, the FDA is currently considering mechanisms by which the postmarket safety reporting requirements ordinarily associated with the marketing application used to approve or clear the combination product may be supplemented, as appropriate, to take into account the combination nature of the product.

Adverse events

Due to a lack of regulatory guidance specific to reporting adverse incidents related to combination products, it is complicated to determine a consistent path for addressing the reporting of adverse events. As a result, this creates significant issues for reporting of investigational adverse events and adverse device effects, as well as adverse drug experiences, adverse device experiences, and MDR reporting, which have different reporting thresholds and timetables.

Some examples of various regulation requirements are listed below:

- The regulations at 21 CFR § 314.80 require expedited reporting of serious adverse drug experiences within 15 calendar days, but only if unexpected.
- In contrast, the device regulations require reporting of an adverse device experience within 30 days if the device may have caused or contributed to a death or serious injury, or has malfunctioned and would be likely to cause or contribute to a death or serious injury if the malfunction were to recur.
- In addition, MDR reportable events that necessitate remedial action to prevent an unreasonable risk of substantial harm must be reported in 5 workdays.

The above differences and uncertainty on which FDA center should receive the report, or whether dual reporting is required, can result in confusion among manufacturers of companies involved with combination products trying to meet these requirements and remain in compliance.

The unspecified regulations for adverse reporting can also lead to inconsistency in compliance and, in many instances, duplication in reporting or overreporting. This can also lead to inconsistency in enforcement. The choice a sponsor makes on what regulations apply to its specific combination product may vary depending on which FDA center takes the lead in the review of the initial product application.

Some examples of conflict in determining the correct reporting structure for adverse events based on the components of a combination product can be seen in the examples provided below:

- When a device or drug is involved in an incident with a patient, it may be unclear to the physician involved which constituent of the combination product was responsible for causing the incident. It would be questionable whether the event occurred as a result of a drug reaction or by a device incident or malfunction. A possibility in this case is that the manufacture's report could result in being submitted to the wrong FDA center, reported out of the required timeframe, or be left unreported totally.

- A reusable insulin pen injector cleared by CDRH under a 510(k) and a disposable prefilled insulin pen injector approved by CDER under a new drug application (NDA) would be subject to different reporting obligations. If serious patient hypoglycemia were to occur in a patient, due to a malfunction of the device, the centers could adopt two interpretations of expedited reporting requirements:
  - CDER, which has primary jurisdiction in the case of the prefilled disposable pen, might conclude under the drug regulations that because (21 CFR 803.50, 805.53) hypoglycemia is an expected adverse event with insulin therapy, the event is nonreportable.
  - CDRH, in contrast, which has primary jurisdiction in the case of the reusable pen, might conclude under the device regulations that the expectancy of the event is irrelevant, and therefore the malfunction is reportable.

Possible options for adverse event reporting

The FDA approval or clearance of a particular constituent part of a combination product is a great initial step for considering the appropriate data for establishing safety and effectiveness for the constituent's use in a combination product. It is recommended by the FDA that developers seriously consider what has already been established about a particular constituent part. For example, consider available data and information to reduce redundancies in duplication and to ensure that an efficient and timely process takes place during product development.

The development team, however, needs to recognize that the combination product being developed does not just constitute separate components, but is a total combined product.

Additional data and information may also be necessary to address specific technical, scientific, regulatory, or clinical issues raised in the development of the novel combination product. These issues may be raised by combining the constituent parts or by new uses and indications requested for the constituent in the combination product.

Combination products approved under one marketing application

One option might be for a combination product approved or cleared under a single marketing application to be subject to the same postmarket safety reporting requirements as those ordinarily applicable to products approved or cleared under the same application type. For example, a combination product that is a drug-device product approved under an NDA would remain subject to the requirements specified in 21 CFR Part 314, as well as device malfunction and 5-day MDR reporting. A device-blood combination product approved under a PMA would be subject to medical device reporting under 21 CFR Part 803, as well as 15-day alert reporting of serious and unexpected adverse events associated with the biologic constituent, and reporting of blood-related deaths, as appropriate.

Combination products approved under separate marketing applications

For some combination products, the constituent parts are approved under separate marketing applications held by the same sponsor. In such a situation, one option would be if the most likely associated constituent part can be determined from the initial safety report obtained from the healthcare provider, hospitals, and so on, the sponsor would submit the safety report as usual required for that particular constituent part. An example can be seen in the case where an adverse event was most likely related to the device constituent of a combination product, in which case an MDR would be submitted.

In a situation where the initial safety report had inadequate or incomplete information to determine which component of the combination product was most likely to be associated with the adverse event, the sponsor would submit the safety report ordinarily received by the lead center for the combination product. An example of this can be demonstrated if CDER was the lead Center for reviewing the combination product, the safety report would be submitted to MedWatch for the drug safety reports.

When the most probable associated constituent part can be determined from the initial safety report (e.g., from a consumer, healthcare provider, or user facility), the sponsor would submit the safety report as ordinarily required for that constituent part. For example, if the event probably was related to the product's device component, an MDR would be submitted. If the initial safety report did not contain enough information to ascertain which constituent part most likely is associated with the event, the sponsor would submit the same type of safety report as ordinarily required by the combination product's lead center. For example, if CDER was the combination product's reviewing center, the safety report would be submitted to the Med Watch mailing address for drug safety reports.

Other combination products, where the constituent parts are approved under separate marketing applications, by different sponsors

In this case, each sponsor would comply with the safety reporting regulations ordinarily associated with the marketing application used to approve or clear its constituent part, for example, medical device, drug, or biologic.

For a sponsor (company A) who receives an adverse event report about a constituent part approved or cleared under the marketing application held by the other sponsor (company B), the one who received the report, that is, company A, would send the report to the other marketing application holder (company B) to assess whether the report should be submitted to the FDA. Company B then maintains records of the initial report and submits the safety report to the FDA with a cover letter indicating the suspect constituent part's manufacturer and application number.

If the initial safety report did not contain adequate information to identify which of the constituent parts may have been associated with the event, the sponsor who received the report submits the report to the FDA, as it would for other reportable events concerning its product, and provides the other constituent part's application holder a copy of the report.

How are adverse events reported for combination products?

As seen above, combination products sometime raise unique concerns about safety and effectiveness, or risks to the public health, arising specifically from the combination nature of the product. As an example, a drug-coated device may be subject to the device quality systems regulation for the device component, to drug good manufacturing practices (GMP) for the drug coating, and to a mix of requirements, as appropriate, for the combined product. Since manufacturers must design their manufacturing and quality systems to address the types of products they produce, a sponsor who primarily manufactures devices may not have the systems in place to manufacture a drug-coated device that will be subject to drug GMP.

Similarly, there is also confusion by product sponsors in deciding which adverse event monitoring regulations to follow for a combination product, and that reporting to multiple centers in some cases is duplicative, cumbersome, and unnecessary.

There is also concern by some sponsors that drug GMP are inadequate to address complex drug delivery systems (some of which may be assigned to CDER), and that such products should instead be subject to the quality systems regulations typically applied to medical devices.

Additional considerations for postmarket requirements for combination products include whether drug versus device advertising and promotion requirements are mandated by the agency, which change system should be used for implementing product modifications, and how new indications for use can be incorporated, for the evolving second-generation combination products.

As another example for reporting adverse events for a combination product, if a product is approved under drug regulations, adverse events associated with the product should be reported under the drug reporting regulations.

Postmarket challenges for combination products

There are significant challenges in the postmarketing of combination products related to the lack of or unspecified requirements in the FDA guidance on the following areas:

- How should one address postmarket changes for medical device constituents of combination products approved under a new drug application (NDA)?

- What are the regulations that are applied to postmarket adverse event reporting requirements (21 CFR 314 or 21 CFR 803)?
- Does a sponsor need to include device investigation results in the expedited device reports?
- In the case where products intended to be used together are cross-labeled, if a device is not cleared to be used for specific products, and in many instances a single drug may be used with several different devices, it is unclear for sponsors what requirements need to be met. There is a strong need for guidance on how the drug label can incorporate these devices and meet any required labeling requirements.
- Additionally, a device manufacturer for a combination product could make changes (postmarket) to the device that may be unknown to its partnering drug company.

International marketing of combination products adds additional requirements and challenges for industry regarding preparation of documents, specific quality system requirements, control and submission of any postapproval changes, regulatory reporting, labeling, and compliance inspections.

Modification of combination products postmarket

There is a strong need for guidance from the FDA to clarify the regulatory pathway for any changes made to constituents of already approved combination products in the market. This can be seen especially with the device component of a combination product that differs from drugs and biologics in the fact that devices usually go through design changes continuously in the course of their evolution for improved performance based on feedback from customers, market performance, and so on. For the device constituent of a combination product, there is still a requirement for a pathway defined by the FDA for clearing and approving these evolutionary changes to the device component in instances where the specific change to the device would not affect the other components. We support the idea of using a medical device design control process to manage change control issues for all types of combination products.

There is also uncertainty in how the FDA would regulate postmarket studies and surveillance for combination products. For example, for a drug-device combination product that is regulated as a device, could the FDA have authority to mandate postmarket studies and surveillance for drug-related safety problems? In the same example, if the combination product is regulated as a drug, would the FDA still have authority to require surveillance of device-related safety problems? Regardless of how this is eventually addressed by the FDA, we believe it is important to address the needs of both the industry and the FDA to avoid under- and overreporting.

Postmarket monitoring

Postmarket monitoring is used to distinguish a company's postdistribution activities from the FDA's mandatory postmarket surveillance program. Postmarket information is used to complement data obtained in premarket clinical studies. A medical device, drug, or biologic combination product in commercial distribution may have different challenges than those that were used in clinical trials.

An example can be seen for companies that may face long-term safety issues, populations with more diverse disease conditions or adjunct physical problems, accidental or deliberate product misuse, and unexpected product malfunctions. Therefore, it is imperative that companies manufacturing combination products review and implement the changes observed from experience and gained from postproduction and implement appropriate corrective actions to correct any identified deficiencies.

Many companies often rely on the complaint handling process to detect current or potential problems. When a problem is detected, the actual root cause needs to be identified, not just to correct the immediate nonconformance, but also to prevent future occurrences. For example, significant changes to the device component of a combination product (i.e., form, fit, or function) must be routed through the design control process. When preparing to make a change, whether it is to the device design, a component, or manufacturing process, the manufacturer should consider the following issues:

- Does the change need to be verified or validated?
- Does the company's change control procedure require this question to be asked, answered, and documented?
- Is there a mechanism to link the corrective action to the design process?

Complaints

Whether a company has a device-biologic or drug combination product, it is responsible for following up on product complaints in the market postlaunch and capturing and resolving these complaints. The various regulations governing drug, device, or biologics provide guidances by the FDA on how to address complaints. For combination product handling, the FDA needs to provide a guidance document on how best to handle such complaints, which combination product component manufacture would play a lead role in reporting the complaint, whether both companies, for example, drug and device for the combination product, need to file complaints, and so on.

Complaints are an extremely useful source of obtaining postmarket feedback from customers, physicians, and so forth, for a company, as it can potentially identify potential product or manufacturing problems to investigate and improve processes or product.

Table 8.1 provides a high-level overview of complaint regulation requirements for device, drug, and biologics. In general, the regulations require

Table 8.1 Complaint Regulation Requirements for Device, Drug, and Biologic Companies

Medical device regulations: 21 CFR 820.198	Drug/biologic regulations: 21 CFR 211.198
Written procedures are required for the following: • Receiving, reviewing, and evaluating complaints (documented and oral) by a designated unit • All complaints to be processed in a uniform and timely manner • Receiving, reviewing, and evaluating complaints (documented and oral) • Oral complaints documented upon receipt • Complaints evaluated to determine whether a medical device reporting (MDR) event has occurred	**Written procedures are required for the following:** • Handling of all written and oral complaints • Review to determine whether the complaint represents a reportable, serious, and unexpected adverse drug experience
Requirements: Written record shall include the following information if available: • Device name • Date complaint was received • Any device identification (I.D.) and control numbers • Name, address, and phone number of complainant • Nature and detail of complaint • Date and result of investigation • If no investigation, reason and name of individual responsible for the decision • Any corrective action taken • Reply to complainant • Complaints that are MDRs must be clearly identified • Complaint files must be available during an FDA inspection	**Requirements:** Written record shall include the following information if available: • Name and strength of the drug product • Lot number • Name of complainant • Nature of complaint • Reply to complainant • Finding of an investigation or the reason an investigation was not necessary • Complaint files must be available during an FDA inspection
Retention time: • All records shall be retained for a period of time equivalent to the design and expected life of the device, but in no case less 2 two years from the date of release for commercial distribution by the manufacturer	**Retention time:** • Written records involving a drug product shall be maintained until at least 1 year after the expiration date of the drug product or 1 year after the date that the complaint was received, whichever is longer

similar requirements for complaints from a biologic, drug, or device company.

Complaints are just one form of feedback. Postmarket monitoring should also include input from sales and marketing personnel, distributor reports, service reports, quality test results, internal audits, clinical investigations, postmarket surveillance studies, medical literature, and regulatory and enforcement activities to assess the true performance of the product in the market. A company needs to be proactive and analyze a specific product or family of products on an annual or semiannual basis. It should not wait for a problem to occur to initiate action. An organization can use the information from the following sources to increase product safety and effectiveness, customer satisfaction, and future market share:

- Evaluate safety alerts, recalls, customer complaints, medical device reports, and review vigilance reports.
- Design and process changes, as well as a summary of product sales and competition and market forecasts.
- Following recalls, medical device reports and warning letters on the FDA website for similar products. Recalls are essential to remove unsafe or ineffective devices from the market. The primary reasons for many recalls are issues such as software, manufacturing/assembly or labeling errors, failure to meet product specifications, potential or real product contamination, sterility problems, design concerns, and component nonconformance. The reasons for several of the subsequent recalls were repeats or incidents similar to the original recalls involving either the same or similar products.
- Did these firms simply correct the specific nonconformance and not eliminate the root cause? If a problem is repeated after corrective action has been implemented, the initial action was most likely ineffective.

A warning letter is issued by the FDA when notifying industry that a violation of regulatory significance exists and that immediate voluntary correction is required. If the issue is not resolved by the company adequately, a continuation of the problem may result in a violative product. As a result, during a subsequent inspection, should an FDA investigator determine that the previous nonconformances were not satisfactorily resolved, the FDA may take enforcement action to protect the public health. There were 420 warning letters issued by the FDA from January 2006 through July 2007. Of these, 11 were issued by CBER, 51 by CDRH, 45 by CDER, 27 by the Center for Food Safety, and 5 by the Center for Veterinary Medicine, with the remaining issued by the district offices.

Most of the problems described in the recalls and warning letters could or should have been identified during routine internal quality audits and thereafter resolved. If adverse actions are repeatedly observed with a particular product type, the FDA is likely to increase facility inspections within that industry.

An example of a warning letter for a biologic facility could potentially include the following nonconformances:

1. "Inadequate investigations" into bioburden excursions relating to five bulk product lots exceeding virus harvest interim bioburden action limits and the "lack of implementation of appropriate corrective and preventive actions, coupled with deficiencies in aseptic practices by personnel, cleaning validation of equipment, and effectiveness of the cleaning and disinfection processes used (http://fda.gov/foi/warning.htm)." Adequate and complete investigations of deviations are required by 501(a) 2(B) of the *Food, Drug, and Cosmetic Act*.
2. Failure to ensure that operators who performed setup, sterile filtration, or aseptic processing "use[d] proper aseptic techniques to prevent microbial contamination of product lots (http://fda.gov/foi/warning.htm)."
3. A "lack of specificity" in master and batch production records.
4. Failure to establish the effectiveness of cleaning and disinfection processes.
5. Cleaning validation of various equipment; the agency noted it had identified microorganisms associated with eggs during environmental monitoring of the facility and its personnel.

Combination product manufacturers must be proactive throughout their product's life cycle. This involves more than evaluating customer complaints, analyzing nonconforming trend reports, reacting to internal audit or regulatory inspectional observations, and establishing a closed-loop corrective and preventive action system. An effective program also includes systematic evaluation of design changes, process changes, device sales, competition, market forecasts, and regulatory trends.

Risk management

Risk management is also a key factor in designing and manufacturing medical devices and combination products, as the risk associated with each constituent, that is, drug-device or biologic-device, needs to be evaluated for risk in the finished product. Risk analysis should therefore be an integral part of each development phase for combination products to identify hazards and harms associated with the product, estimate and evaluate the associated risks, implement risk control measures, and monitor the effectiveness of the measures. Note that risk management must be managed throughout the product design process and life cycle (e.g., as per ISO 14971).

Risk management, however, does not stop after products have been released. Companies need to actively follow their product performance over time in the market and collect data on significant risks to patients by use of the product, and ultimately negate these risks by addressing them.

Key considerations in an example where a drug or biologic is added to a device to develop a combination product are shown below:

- Medical device companies should apply their risk management procedures and approaches to combination product development to ensure that all potential hazards associated with the combination are appropriately analyzed.
- The product development teams conducting risk management should be expanded to include cross-functional teams from appropriate disciplines from all partners involved in the development and manufacture of the final combination product.
- Risk management practices should occur throughout the life of the product, and especially after market launch.

Table 8.2 lists certain key activities to be considered for postmarket launch of combination products.

In conclusion, some of the postapproval issues faced by companies that have launched combination products are even more challenging since substantial

Table 8.2 Activities for Consideration for Postmarket Launch.

Task no.	Activity
1.	Implement the postmarket surveillance plan.
2.	a. Track postlaunch marketing activity and metrics. b. Conduct primary market research to obtain feedback on customer needs and perceived product performance. c. Identify postmarket clinical activities and product improvement feedback.
3.	a. Track postlaunch operation's activity and metrics such as cost of goods sold, process yields, capacity plans, etc.
4.	a. Track postlaunch quality activity and metrics for complaints, adverse events, compliance audits, etc. b. Based on data obtained, update the risk analysis as necessary. c. Situations that may require additional information if there are new or expanded conditions for use for an existing product, or a need for longer-term follow-up or evaluation of rare/unusual events.
5.	If postlaunch significant issues are observed, the team needs to determine and implement root causes for the effects of the product in the market and address these with a permanent corrective action and preventative action (CAPA), or design changes in the next-generation product.

information is available on the preapproval and market clearance processes, but fails to address postapproval issues adequately in how the product will be regulated once approved. As in the case of prelaunch development efforts, we encourage companies to carefully plan, analyze each situation based on its own merit, rigorously execute, and follow up on postlaunch efforts for combination products.

Bibliography

21 CFR, Part 314. *Drug products are subject to postmarket reporting*, 2006.

21 CFR, Part 803. *Medical device reporting* (*MDR*), 2006.

21 CFR, Part 600. *Biological products are subject to vaccine adverse event reporting system* (*VAERS*), 2006.

Adoption of Quality Systems Model for Agency Operation. Office of Combination Products, November 2004.

Adverse event reporting and number of marketing applications for combination products: http://www.fda.gov/oc/combination/reqcomm905.html

Application of cGMP Regulations to Combination Products: frequently asked questions. Guidance for industry and staff, Office of Combination Products, January 15, 2007.

Balboni, Mark L. Post marketing application integrity: what is data integrity and how to ensure it? Regulatory Affairs Focus, April 2006.

Comments on FDA adverse event. Concept paper, Combination Product Coalition (CPC), March 23, 2006.

Critical path opportunities report. *Federal Register* 06-354, January 17, 2006.

FDA database website: http://www.accessdata.fda.gov

IND meetings for human drugs and biologics, chemistry, manufacturing, and controls information guidance for industry: http://www.fda.gov/cder/guidance/3683fnl.pdf

Keramidas, Sherry. The challenge of combination products and beyond, Regulatory Affairs Focus, September 2006.

Klassen, Dan G. Post marketing/GMPs, after licensure no time to relax. Regulatory Affairs Focus, April 2006.

Leichter, Lee. Medical device combination products. Regulatory Affairs Focus, February 2001.

Medical Device Innovation Initiative. FDA, CDRH, May 2006.

Modifications to devices subject to premarket approval. Draft guidance for industry and FDA staff. *Federal Register* (72)58:14282–14283, March 27, 2007.

Pharmaceutical cGMPs for the 21st century—a risk-based approach. Office of Combination Products, 2003.

Postmarket safety reporting for combination products. FDA concept paper, September 2005: http://www.fda.gov/ocp.com

Smith, Barbara-Helene. Post marketing/GMPs, postmarket medical device monitoring program. Regulatory Affairs Focus, April 2006.

Van Buskirk, Gale E. A GMP-compliant quality system for a combination product. Regulatory Affairs Focus, pp. 37–39, January 2005.

chapter nine

Agency audits and challenges

Chapter 9 focuses on the Food and Drug Administration (FDA) agency audits in regards to biologic, pharmaceutical, and medical device industries. The initial sections of this chapter focus on the traditional roles played by the CBER, CDER, and CDRH divisions within the FDA, and how the agency audits and the types of audits specific to each branch have been traditionally managed. This is followed by a section that focuses on inspections of manufacturers of combination products and some of the challenges these manufacturers face at present.

The medical device, pharmaceutical, and biologic industries are all highly regulated in the world, and this is especially true within the United States. The usual method for these companies to be compliant to the required regulations is by developing, maintaining, and implementing documented internal processes, policies, and procedures that meet the applicable regulatory requirements and directives. The implementation of a strong compliance program within any given company usually stems from a clear vision, commitment, and expectations set forth by upper management. In many instances the start-up or maintenance cost of a strong compliance program seems daunting. However, the rewards and benefits from such a strong program usually result in greater business rewards and less risk. The risks for any company to be noncompliant can be costly not only financially, but also in reputation, FDA agency outcomes during inspections, field performance, and customer perception of the company and its products.

Postmarket monitoring

As discussed in Chapter 8, areas such as risk management are key factors in designing and manufacturing medical devices, and risk analysis should be an integral part of each development phase. Risk management, however, does not stop after products have been released. Companies need to actively monitor their devices' performance. The term *monitoring* is used to distinguish a firm's postdistribution activities from the FDA's mandatory postmarket surveillance program. Postmarket information complements data collected in premarket clinical studies. Devices in commercial distribution may have different challenges than those used in clinical trials. For example, products can exhibit long-term safety issues due to populations with more diverse disease conditions or adjunct physical problems, accidental or deliberate device misuse, and unexpected product malfunctions. Therefore, it is imperative that medical device firms review experience gained from postproduction devices and implement appropriate corrective action.

Most companies rely upon the complaint-handling process to detect current or potential problems. When a problem is detected, the true root cause needs to be identified—not just to correct the immediate nonconformance, but also to prevent future occurrences. Changes can result from actions to correct issues. Significant changes to the device, that is, form, fit, or function changes, must be routed through the design control process. When preparing to make a change, whether it is to the device design, a component, or manufacturing process, the manufacturer should consider asking questions such as:

- Does the change need to be verified or validated?
- Does the firm's change control procedure require this question to be asked, answered, and documented?
- Is there a mechanism to link the corrective action to the design process?

Complaints are just one form of feedback. Postmarket monitoring also should include input from sales and marketing personnel, distributor reports, service reports, quality test results, internal audits, clinical investigations, postmarket surveillance studies, medical literature, and regulatory and enforcement activities. A company should not wait for a problem to occur in order to initiate action. Being proactive and analyzing a specific product or family of devices on an annual or semiannual basis may be a good opportunity to assess the status of specific product families. Evaluating safety alerts, recalls, customer complaints, medical device reports, vigilance reports, and design and process changes, as well as a summary of device sales and competition and market forecasts, may be another means of assessing the status of current systems, documentation, and processes for these products. The information thus obtained can then be utilized to increase device safety and effectiveness, customer satisfaction, and future market share.

Checking recalls, medical device reports, and warning letters on the FDA website for similar products should be a key for organizations to assess competitor performance in the market, as well as obtaining information on new technological products. A warning letter is the FDA's principal means of notifying industry that a violation of regulatory significance exists and that prompt voluntary correction is required. The warning letter establishes prior notice. If the company does not permanently resolve the issue, a continuing problem may result in a nonconforming product. If, during a subsequent inspection, an FDA investigator determines that a previous nonconformance had been unsatisfactorily resolved, the agency can take enforcement action to protect the public.

Many of the problems described in recalls and warning letters can usually be identified by an organization's internal quality system–related activities, such as routine internal quality audits, complaints monitoring, and so forth, and resolved prior to an agency audit. It is safe to assume that the FDA is probably going to have increased facility inspections within that industry if adverse actions are repeatedly observed with a particular product type within a firm or a given industry to ensure product safety and product integrity.

FDA's post-approval audit program

The FDA's *Compliance Program Guidance Manual: Post approval Audit Inspections* (1994) outlines the FDA's post-approval inspection process for drug, device, and biologic products. Through the post-approval audits of approved products, whether a new drug application (NDA) or an abbreviated new drug application (ANDA), the FDA confirms that postmarketing compliance is met by approved manufacturing establishments to ensure the following expectations:

1. Any changes in manufacturing and process control are in compliance with GMP regulations.
2. All changes are documented in either supplemental applications or annual reports.
3. Confirmation that requirements for adverse reaction reports, NDA field alerts, and annual reports are being met, and verification that commitments made by companies at the time of application approval are completed or under way, in accordance with the agreement.

Traditional FDA inspections by agency

CBER: Biological products

Center for Biologics Evaluation and Research (CBER) responsibilities include a wide range of regulatory activities during the life cycle of biological products. Manufacturing facility inspections are conducted to assess whether biological products are made in compliance with the appropriate laws and regulations. Inspections are conducted prior to products being licensed or after certain changes are made and approved. In most cases, routine GMP inspections are conducted periodically to ensure continued compliance. An integrated approach is used by CBER for the managed review of licensed biological products by the agency performing the preapproval/prelicense inspections (PAIs/PLIs). This is done as part of the complete review of the marketing submissions. CBER staff conduct these inspections.

CBER's Bioresearch Monitoring (BIMO) program also includes on-site inspections of clinical investigators, sponsors/monitors, nonclinical testing labs for good laboratory practices, and institutional review boards (IRBs). These inspections are performed to assess the safety and protection of the rights and welfare of human research subjects, and verify the quality and integrity of data submitted to CBER. Inspections are also conducted to investigate complaints received from subjects who participated in such clinical trials. CBER continues to monitor the safety, purity, and potency of licensed biological products after approval, and its monitoring activities include the following:

1. Routine GMP inspections of CBER's licensed biological drugs and devices are performed by team biologics with participation from CBER product specialists.
2. Investigators from the FDA's Office of Regulatory Affairs (ORA) conduct routine inspections of establishments that manufacture blood and blood products, source plasma, human cells and tissues, and cellular and tissue-based products (HCT/Ps) according to established compliance programs.
3. Analysis of biological product deviation reports (BPDRs) and HCT/P deviation reports, which can provide an early warning of problems in manufacturing. Manufacturers are required to report certain deviations in manufacturing of distributed licensed biological products, blood components, and HCT/Ps. Information from BPDRs is used to identify possible recall situations.
4. CBER personnel monitor investigations into transfusion- and donation-related fatalities and complaints from consumers and other sources.
5. In addition to required reporting by manufacturers, health professionals and members of the public are encouraged to report adverse events or reactions with biological products to the FDA's MedWatch and, for vaccines, to the Vaccine Adverse Event Reporting System (VAERS).

CBER personnel are responsible for the evaluation of the health hazard associated with biological products that are voluntarily recalled or withdrawn from the market by manufacturers. When a manufacturer is not in compliance with applicable laws and regulations, the FDA can take a variety of regulatory actions. Advisory actions inform the manufacturer of violations that should be promptly corrected. Administrative or judicial actions may be considered when violations represent a significant public health risk.

The actions by the agency can include suspension or revocation of biologics licenses or civil and criminal penalties. When a clinical investigator is conducting studies of biologics products in a manner that poses an undue risk to safety of the human subjects, CBER can take specific enforcement actions, such as disqualification or debarment. CBER also monitors the Internet and advises website owners that selling unlicensed or unapproved products in the United States is a prohibited act.

CDER: Drug products

The Center for Drug Evaluation and Research (CDER) branch of FDA, by carefully monitoring pharmaceutical manufacturing facilities for compliance with its current good manufacturing practice (cGMP) regulations, tries to ensure that the quality of drug products marketed is safe for the public. The cGMP regulations for drugs contain minimum requirements for the methods, facilities, and controls used in manufacturing, processing, and packaging of a drug product. The regulations make sure that a product is safe for use, and that it has the ingredients and strength it claims to have. The approval process for new drug and generic drug marketing applications includes a review of the manufacturer's compliance with the cGMP. FDA inspectors determine whether the firm has the necessary facilities, equipment, and skills to manufacture the new drug for which it has applied for approval. Decisions regarding compliance with cGMP regulations are based upon inspection of the facilities, sample analyses, and compliance history of a company.

The FDA can issue a warning letter or initiate other regulatory actions against a company that fails to comply with current good manufacturing practice regulations. Failure to comply can also lead to a decision by the FDA not to approve an application to market a drug. With the recent focus on quality by design (QbD), process analytical technology (PAT), risk management, and so on, it is only a matter of time before FDA inspectors begin to review these in addition to a manufacturer's compliance with the GMP prior to product approval.

The Division of Compliance Risk Management and Surveillance

The Division of Compliance Risk Management and Surveillance is responsible for the identification, assessment, and prioritization of legal violations based on their public health significance. The division obtains this by use of qualitative or quantitative data analysis and strategic problem solving to target compliance actions that are used to develop enforcement strategies for reducing public health risks associated with drug product problems.

The Division of Compliance Risk Management and Surveillance is responsible for some of the activities listed below:

- Ensuring drug and patient safety by enforcing the postmarketing adverse drug experience (ADE) reporting regulations to ensure timely and accurate submission of ADE reports on adverse drug reactions.
- Supporting CDER's Office of Drug Safety (ODS) with postmarketing drug safety surveillance operations.
- Monitoring the quality of the nation's drug supply through postmarked surveillance sample collection and analysis, the drug quality reporting system, and field alert reports.
- Identifying less-than-effective drug products in support of the Medicaid Drug Rebate Program.

Inspections for drug manufacturers

The primary mission of the Food and Drug Administration is to conduct comprehensive regulatory coverage of all aspects of production and distribution of drugs and drug products. The FDA has developed two basic strategies to carry out these functions:

1. The FDA performs evaluation through factory inspections, which includes the collection and analysis of associated samples, and the conditions and practices under which drugs and drug products are manufactured, packed, tested, and contained.
2. The FDA also monitors the quality of drugs and drug products through surveillance activities such as sampling and analyzing products in distribution.

The goal of the above activities is to minimize consumer exposure to adulterated drug products.

The FDA performs the inspections for the reasons listed below:

1. To determine whether inspected companies are operating in compliance with their applicable cGMP requirements, and if not, can these companies provide evidence for actions taken to prevent adulterated products from entering the market and, if needed, to remove adulterated products from the market
2. To provide cGMP assessment that may be used to determine the acceptability of the company in the preapproval review of a facility for new drug applications
3. To provide input to companies during inspections to improve their compliance with regulations
4. To continue the FDA's unique expertise in drug manufacturing in determining the adequacy of cGMP requirements, agency cGMP regulatory policy, and guidance documents

There are different types of inspections performed for drug companies. One type, biennial inspection of manufacturing sites, includes repackaging, contract labs, and so on. Biennial inspections (every 2 years) are conducted under this program to:

1. Reduce the risk that adulterated products are reaching the marketplace
2. Increase communication between the industry and the agency
3. Provide for timely evaluation of new manufacturing operations in the company
4. Provide for regular feedback from the agency to individual companies on the continuing status of the company's GMP compliance

The inspections would focus on the following systems for drug manufacturers:

1. ***Quality system:*** Should ensure overall compliance with cGMP and internal procedures and specifications. The system should include the quality control unit and all of its review and approval duties (such as change control, reprocessing, batch release, annual record review, validation protocols, reports, etc.). It also needs to include all product defect evaluations and evaluation of returned and salvaged drug products (cGMP regulation, 21 CFR 211 Subparts B, E, F, G, I, J, and K).
2. ***Facilities and equipment system:*** Should include the measures and activities that provide an appropriate physical environment and resources used in the production of the drugs or drug products. It includes:
   a. Buildings and facilities along with maintenance.
   b. Equipment qualifications (installation and operation), equipment calibration and preventative maintenance, and cleaning and validation of cleaning processes as appropriate. Process performance qualification will be evaluated as part of the inspection of the overall process validation that is done within the system where the process is employed.
   c. Utilities that are not intended to be incorporated into the product, such as HVAC, compressed gases, and steam and water systems (cGMP regulation, 21 CFR 211 Subparts B, C, D, and J).
3. ***Materials system:*** Should include measures and activities to control finished products and components, including water or gases that are incorporated into the product, containers, and closures. It includes validation of computerized inventory control processes, drug storage, distribution controls, and records (cGMP regulation, 21 CFR 211 Subparts B, E, H, and J).
4. ***Production system:*** Should include measures and activities to control the manufacture of drugs and drug products, including batch compounding, dosage form production, in-process sampling and testing, and process validation. It also includes establishing, following, and documenting performance of approved manufacturing procedures (cGMP regulation, 21 CFR 211 Subparts B, F, and J).
5. ***Packaging and labeling system:*** Should include measures and activities that control the packaging and labeling of drugs and drug products. It includes written procedures, label examination and usage, label storage and issuance, packaging and labeling operations controls, and validation of these operations (cGMP regulation, 21 CFR 211 Subparts B, G, and J).
6. ***Laboratory control system:*** Should include measures and activities related to laboratory procedures, testing, analytical methods

development and validation or verification, and the stability program (cGMP regulation, 21 CFR 211 Subparts B, I, J, and K).

Options for surveillance inspections

There are several options for the surveillance inspections, such as (1) the full inspection and (2) an abbreviated inspection, which are described below.

The full inspection option The full inspection option is a surveillance or compliance inspection that is meant to provide a broad and deep evaluation of the company's cGMP. This will be performed when little to no information is known about a particular company's cGMP compliance, for example, in the case of new companies, or for companies where there is doubt about the cGMP compliance state within the company, or follow-up to previous regulatory actions. Based on findings of objectionable conditions per one or more systems (a minimum of two systems must be completed), as discussed in the section above, a full inspection may revert to the abbreviated inspection option, with district concurrence.

During the course of a full Inspection, verification of quality system activities may require limited coverage in other systems. The full inspection option would normally include an inspection audit of at least four of the systems, one of which must be the quality system (the system that includes the responsibility for the annual product reviews).

The abbreviated inspection option The abbreviated inspection option is a surveillance or compliance inspection that is meant to provide an efficient update evaluation of a company's cGMP status. The abbreviated inspection would provide documentation for continuing a company in a satisfactory cGMP compliance status. Generally this would be done when a company has a record of satisfactory cGMP compliance, with no significant recalls or product defect or alert incidents, or with little shift in the manufacturing profiles of the company within the previous 2 years. The abbreviated inspection option normally would include an inspection audit of at least two of the systems, one of which must be the quality system (the system that includes the responsibility for the annual product reviews).

Compliance inspections

Compliance inspections are performed to evaluate or verify compliance corrective actions after a regulatory action has been taken. First, the coverage given in compliance inspections must be related to the areas found deficient and subjected to corrective actions.

In addition, coverage must be given to systems because a determination must be made on the overall compliance status of the company after the corrective actions are taken. The company is expected to address all of its operations in its corrective action plan after a previously violative inspection, not just the deficiencies noted in FDA-483.

Compliance inspections include for-cause inspections. For-cause inspections are compliance inspections that are done to investigate a specific problem that has come to the attention of some level of the agency. The problems may be indicated in field alert reports (FARs), industry complaints, recalls, indicators of defective products, and so on. Coverage of these areas may be assigned under other compliance programs. However, expansion of the coverage to a GMP inspection is to be reported under this program.

CDRH: medical devices

The Center for Device and Radiological Health (CDRH) compliance program by the FDA for medical devices focuses on administrative/enforcement activities related to the quality system (QS) regulation (21 CFR Part 820), the medical device reporting (MDR) regulation (21 CFR Part 803), the medical device tracking regulation (21 CFR Part 821), the corrections and removals regulation (21 CFR Part 806), and the registration and listing regulation (21 CFR Part 807). The FDA's compliance program for medical devices encompasses five regulations for inspecting medical device firms:

1. Under the quality system regulation, manufacturers are expected to control their devices from the design stage through postmarket surveillance.
2. Manufacturing processes, such as sterilization, are required to be implemented under appropriate controls.
3. MDR tracking.
4. Corrections and removals regulations involve activities with which manufacturers and importers are required to comply after the devices are distributed.
5. Registration and listing regulation.

We will now provide additional details on these five regulations.

The quality system (QS) regulation

Manufacturers are expected to establish and follow quality systems to help ensure that their products consistently meet applicable requirements and specifications. The quality systems for FDA-regulated products (biologics, drugs, and devices) are known as cGMP. cGMP requirements for devices (in 21 CFR Part 820) were first authorized by section 520(f) of the federal *Food, Drug, and Cosmetic Act*. Under section 520(f) of the act, the FDA issued a final rule in the *Federal Register* of July 21, 1978 (43 FR 31508), prescribing cGMP requirements for the methods used in, and the facilities and controls used for, the manufacture, packing, storage, and installation of medical devices. This regulation became effective on December 18, 1978.

The Safe Medical Devices Act of 1990 (SMDA), enacted on November 28, 1990, amended section 520(f) of the act, providing the FDA with the authority to add preproduction design controls to the cGMP regulation. This change in law was based on findings that a significant proportion of device recalls was attributed to faulty design of the product. The SMDA also added section 803 to the act (21 USC 383), which, among other things, encourages the FDA to work with foreign countries toward mutual recognition of cGMP requirements. The FDA published the revised cGMP requirements in the final rule, entitled "Quality System Regulation," in the *Federal Register* of October 7, 1996 (61 FR 52602). This regulation became effective on June 1, 1997, and remains in effect.

The MDR regulation

The first medical device reporting (MDR) regulation became final on December 13, 1984. As a result of changes mandated by the Safe Medical Devices Act (SMDA) of 1990, and the Medical Device Amendments of 1992, the 1984 MDR regulations (21 CFR 803 and 807) were revised and published on December 11, 1995. The FDA Modernization Act of 1997 made additional changes, and a revised MDR regulation was proposed in May 1998. The final revised MDR regulation was published in the *Federal Register* of January 26, 2000. This latest version of MDR regulation includes reporting requirements for manufacturers, user facilities, and importers. MDR reporting for medical device distributors (except importers) was revoked by the FDA Modernization Act of 1997. Distributors are, however, still required to maintain complaint records, per 21 CFR 803.18(d)(1–3).

21 CFR Part 803 requires manufacturers of medical devices, including *in vitro* diagnostic devices, to report to the FDA whenever the manufacturer or importer receives or otherwise becomes aware of information that reasonably suggests that one of its marketed devices:

1. May have caused or contributed to a death or serious injury
2. Has malfunctioned, and that the device or any other device marketed by the manufacturer or importer would be likely to cause or contribute to a death or serious injury if the malfunction were to recur

Note: Importers (initial distributors) of medical devices are subject to 21 CFR Part 803, published in the *Federal Register* of January 26, 2000, and effective March 27, 2000.

The medical device tracking regulation

Under the authority of section 519(e) of the act, the agency may issue a written tracking order that tells a manufacturer to implement a tracking program that meets the requirements of 21 CFR Part 821. Devices subject to tracking may include those that are permanently implanted or life-sustaining/life-supporting devices that are used outside a device user facility. These devices are considered reasonably likely to cause serious adverse health consequences if they fail. The regulation is intended to ensure that in the event of a recall or safety alert, the manufacturer from the device manufacturing facility to the end user or patient can trace a tracked device.

The corrections and removal regulation

The corrections and removal regulation requires manufacturers and importers to report promptly to the FDA any corrections or removals of devices undertaken to reduce risk to health.

The registration and listing regulation

The registration and listing regulation requires manufacturers and foreign exporters to register and list their devices, and importers to register.

This compliance program is used to conduct quality system inspections of devices.

Inspections of medical device manufacturers

Priorities for QS inspections for medical device companies are targeted to cover manufacturers of class II and class III devices, utilizing a risk-based methodology.

The risk-based model below is utilized:

1. Premarket and preclearance inspections under MDUFMA (inspections of manufacturers of devices with a pending PMA approval will be assigned under the PMA compliance program)
2. Manufacturers of class III devices that have never been inspected
3. Compliance follow-up/for-cause inspections
4. Manufacturers of high-risk devices, which can be identified by:
   a. Special assignment from CDRH.
   b. Devices with a higher frequency of recalls and MDRs.
   c. Devices that are driven by software and those with rapidly evolving technological changes. Both of these types of devices are subject to rapid and potentially poorly controlled modifications that could affect their continued safety and efficacy.
   d. New devices that have not been manufactured and distributed for very long.
5. Single-use device reprocessors: Hospital reprocessors and third-party reprocessors

Highest priority should be given to MDUFMA assignments and those class III device manufacturers that have not been previously inspected. The high-risk device category noted in 4 above lists suggestions to the field on how to identify firms for surveillance inspections based on a risk model.

All class I devices, including those exempted from most of the quality system regulation requirements, need to comply with record-keeping requirements and complaint file requirements, including reporting requirements under the MDR regulation. Class I manufacturers should receive lowest inspectional priority unless addressed by a special "for cause" assignment or when a health hazard is apparent.

If inspecting a manufacturer that was originally planned as a class I QS nonexempt, class II or III device firm, and the inspection finds that the firm no longer makes class I QS nonexempt, class II or class III devices, the investigator should review the firm's complaint handling system and MDR practices, then terminate the inspection.

FDA inspections and combination product manufacturers

It is clear from the previous sections that the FDA has established a fairly comprehensive set of regulations to govern the approval, marketing, postmarket surveillance, and so on, of medical devices, pharmaceuticals, and biologics. Companies that are developing and manufacturing combination products have additional requirements that have to be met to avoid noncompliance issues and market products successfully. Compliance issues for combination product companies can occur during development, manufacturing, or postlaunch in terms of advertising or promotional activities.

We will now present two examples of these additional requirements. The first example is where company X in the manufacture of a drug-coated device is expected to store the product at a specific storage condition (refrigerated) due to the degradation of the drug component at higher temperatures. If the warehouse storage or manufacturing operations were not tightly controlled and monitored, the product may not perform adequately once in the market. As a result, a product recall may need to be initiated by the company, which could be costly and leave the patients and market with inadequate availability of product. If this product were a pure device, this requirement may not have been applicable to this company.

A second example can be where a device and biologic product has to be sterilized as a final product. In this case, if the sterilization cycle was compromised, the biologic component of the combination product could degrade, which subsequently could result in the effectiveness of the combination product being compromised in the field. If adverse events were to result, this would also potentially result in product recall. If the product risk is high in the market, the FDA would stop commerce of the product entirely unless the deficiency with the product was corrected. This could result in large losses in sales for a given company.

In many instances, such recalls would very likely result in FDA inspections of a company's manufacturing and other facilities to determine the state of compliance. These inspections could very well go beyond the manufacturing of the product and may include the design process, management responsibility, and so on.

FDA inspections

FDA inspections before the advent of combination products have been managed by specific divisions within the FDA, such as CDER for drugs, CDRH for devices, and CBER for biologics. The process of inspections specific to each office is described in some detail in the later sections of this chapter.

Agency audits for combination products

For many combination product manufacturers, a real concern is the ability to meet regulatory compliance, where more than one set of manufacturing practices may be required for the combination product, as described in Chapters 3 and 4. As a result, manufacturers of combination products try to comply with both device and drug regulations for a drug-device combination product, or biologic and device requirements for a biologic-device combination product. These manufacturers would also benefit from implementing robust quality and manufacturing systems within their organizations that would not only meet the FDA requirements from a compliance standpoint, but also ensure high-quality products that are safe for patient use in the field.

One of the key challenges for the manufacturers of combination products is to understand and implement how and which specific regulations from two or more sets of regulations apply to them. On one hand, if this is done properly, it would result in an optimum approach that will allow these companies to be compliant. On the other hand, if this is not done properly, there is a potential that the companies can go overboard with compliance that may "choke" the design, development, manufacturing, and marketing processes. From an FDA perspective, new policies and regulations certainly take time to draft and implement for industry. The FDA is currently in the process of writing and obtaining feedback on a proposed rule on combination product good manufacturing practices in 2007.

1. In regards to the current good manufacturing practices (cGMP), a company must ensure that the product continues to be manufactured in accordance with GMP regulations, pursuant to 21 CFR 210/211, 606, and 820. Routine inspection by both the FDA and any state or local food and drug authorities would be forthcoming, and the company would need to follow those quality systems that have been put in place and implemented for the past several years. A robust and effective quality system would keep the company out of trouble and prevent many hardships, expenses, and marketing issues with the product.
2. Application/data integrity is also very critical, since a lot of information was compiled regarding processes, methodologies, equipment, utilities, and facilities, and entered into the application to obtain the product license. The company needs to make sure that the information presented in the license always reflects what is happening on site

in the manufacturing and testing arenas. This does not mean that changes cannot be made; however, it does mean that when changes are planned, license supplements need to be submitted to the FDA. It is also necessary for an organization to anticipate when product made using the change may be distributed. The FDA refers to these changes as "changes to an approved application," which are commonly referred to as postapproval changes that usually may arise as a result of scale-up or other customer feedback mechanisms. There are a number of reporting categories, depending upon the change's potential impact to the product's safety, purity, potency, and efficacy, and conditions associated with each change category dictate when product manufactured using that change may be distributed. Some modifications require FDA approval before product made under the change may be distributed. Other, less critical alterations may only require 30 days' lead time for the FDA's initial assessment of the submitted change, or notification either immediately or in an annual update. Although some changes that do not require FDA preapproval allow product distribution before the FDA has completed a technical review of the change report submission, that product is still subject to a market hold or withdrawal should the agency find the submission lacking. In many instances, companies have got into trouble, often during an inspection, when changes have been made that were properly qualified and justified, but not reflected in an update to the product license. This can potentially result in marketing a product that has been manufactured utilizing an unapproved process, design, components, procedures, and so on. Additionally, changes in equipment or processes due to regulatory issues, supply problems, and increase in manufacturing volume or sales could also result in nonconformances if handled inappropriately.

3. It is important to note that the FDA requires a number of reports after licensure. Some are stipulated in the approval letter and others are in the regulations. These include, but are not limited to, the following reports expected by the agency:
   - 21 CFR 314.81, 314.98, 601.12, 814.84: Annual update reports
   - 21 CFR 314.80, 314.98, 600.80, 601.44, 640.73, 803: Adverse events, and so on (drugs, biologics, vaccines, device use facility/manufacturing)
   - 21 CFR 314.81, 600.81: Distribution reports
   - 21 CFR 314.81, 600.14: Field alerts and biological deviation and error reports
   - 21 CFR 314.81, 601.70: Status reports of postmarketing studies
   - 21 CFR 314.81: Advertising and promotional labeling

For all of the post-marketing and GMP information noted above, the FDA has an assortment of final and draft guidances available on its website. In addition, FDA product managers are often a useful resource when it comes to determining exactly what a particular product or situation requires. Companies are often hesitant to ask the agency about such issues, perhaps fearing that a future inspection will then involve looking into those areas. While we understand this predicament, we strongly believe that, considering the alternative, it is undoubtedly in the company's best interests to address potential issues immediately and directly, rather than being concerned about their information sharing.

Medical device manufacturers need to be proactive throughout their product's entire life cycle. This involves more than evaluating customer complaints, analyzing nonconforming trend reports, reacting to internal audit or regulatory inspectional observations, and establishing a closed-loop corrective and preventive action system. An effective program would also need to include systematic evaluation of design changes, process changes, device sales, competition, market forecasts, and regulatory trends.

Various types of FDA inspections

This section provides an overview on the various types of FDA inspections companies may expect during the various stages of the life cycle of a product within a regulated facility.

Pre-approval inspections

For combination products, the FDA often faces challenges in terms of determining the regulatory jurisdiction over the combination product, inspectional authority and site readiness, appropriate preclinical and clinical trial design issues, and establishing appropriate postmarket studies and surveillance. Manufacturers of combination products should ensure that their validations are finalized before the FDA's pre-approval inspections to result in successful outcomes.

Examples of potential, preclinical deficiencies in the case of a drug-eluting stent (device-drug combination products) are represented below:

- There may be a case of inadequate stent platform testing in regards to evaluation of fatigue and corrosion testing. During the addition of a coating to the stent, it is critical to evaluate the fatigue and cracking effects. Additionally, effects of corrosion through potential cracks in the coating need to be evaluated for performance of the finished product.
- Inadequate analysis of any surface modifications made to the device, through application of the coating with the drug substance in it, could have severe consequences to the coating integrity and durability testing of the product.
- Characterization of drug content and its uniformity along the length of the stent is also critical for optimal performance of the finished coated stent.

Post-approval or surveillance inspections

The FDA typically inspects manufacturing sites, animal testing sites, and select clinical study centers. These inspections are usually performed around the time that the marketing application is submitted, or during the subsequent review period.

The FDA works closely with a sponsor to schedule the necessary manufacturing, animal, and study center inspections. FDA district offices may send one or more inspectors qualified in evaluating drug and device quality and manufacturing compliance. Although the sciences of manufacturing drugs and devices are very different, the quality control measures and documentation procedures are similar. This facilitates a relatively straightforward inspection of combination product manufacturing facilities.

Some of the uncertainties that arise for combination product manufacturers in regards to FDA inspections for these facilities include the following questions:

1. How are FDA inspectors alerted to the fact that they are going to be inspecting a combination product manufacturer?
   - If CDER is the lead center involved with the combination product application, the field force would be sent a copy of the chemistry manufacturing and control (CMC) section of a new drug application (NDA) when it is submitted to FDA, or other documentation that would indicate that a combination product manufacturer is being inspected.
   - If CDRH is the lead center, the process will ensure that field force inspecting manufacturers for which CDRH is the lead center will receive copies of a PMA, 510(k) submission, or other documentation letting the inspectors know that they are inspecting a combination product manufacturer.
2. Would only certain inspectors or types of inspectors inspect combination product manufacturers?
   - The FDA's perspective suggests inspectors that are charged with responsibility for inspecting combination product manufacturers would be cross-trained on applicable GMP regulations. Such inspectors would also be trained on the unique issues that face combination products. Additionally, the FDA endeavors to coordinate its inspections such that a team of two or more inspectors with complementary knowledge, skills, and experiences, when appropriate, may inspect combination product manufacturers.
3. How does the FDA ensure that inspectors are appropriately trained?
   - It is suggested that the FDA would ensure that inspectors who inspect combination product manufacturers would be cross-trained on relevant regulations and on combination product–specific issues.

4. If different sets of cGMP regulations apply to a combination product, would the lead center responsible oversee the enforcement of all of those regulations?
    - The FDA suggests that the lead center would oversee and have ultimate responsibility for the inspections. However, the lead center must consult with other centers as appropriate.
5. How would assignment of a lead center affect assignment of inspection personnel for inspections at a combination product manufacturer facility?
    - It is suggested that FDA personnel from the lead center assigned to the combination product would conduct or lead the inspection of a combination product manufacturer. Inspectors who inspect combination product manufacturers will have been trained on relevant regulations. For example, inspectors from CDRH will have been trained on relevant drug and biologic cGMP regulations in addition to the device regulations. The lead center is also charged with consulting with other centers as necessary. An example can be to obtain input on a unique regulatory issue or to coordinate an inspection.
    - In some cases, it may be necessary to have a team of inspectors comprised of inspectors from different FDA centers, for which the lead center needs to determine and coordinate the inspection.
6. How would the FDA determine if a cross-center team of inspectors or separate inspections would be needed to inspect a combination product manufacturer?
    - It is suggested by the FDA that it would determine that inspectors from more than one center are needed to inspect a combination product manufacturer. Many factors could play into this determination; examples of such factors may include:
        - Complexity of the combination product
        - Familiarity of the investigators with the product and the manufacturer
        - Investigators' experience with inspecting combination product manufacturers generally
        - Length of time since previous inspection, and whether a particular manufacturer or product line has ever been inspected before
        - Manufacturer's recommendation
        - Availability of inspectional resources

The FDA lead center would evaluate these factors and determine the type of inspection that is warranted and would oversee the inspection. The following example demonstrates the roles of the various FDA counterparts in determination of product performance.

Drug-eluting stents example

CDRH, as lead center for the drug-eluted stents, conducted the manufacturing inspections for drug-eluting stents; however, reviewers from CDER's Office of New Drug Chemistry were also involved. It is important to note that for these products, the drug regulations (cGMP) had been applied to the manufacturing of the drug substance, and the device quality system regulations (QSR) to the finished product.

Manufacturers of drug-eluting stents should have their validations complete prior to the FDA's preapproval inspection. The FDA works interactively with staff from the two centers and with manufacturers to try to get inspections done as quickly as possible, but it is very important to have all of those validations done to ensure a successful outcome. If there are subsequent changes to the manufacturing process, which would typically require subsequent validations, the FDA may have to go out for a second time to perform follow-up inspections. It is the best use of the agency's resources, as well as for an organization, if inspections are completed successfully the first time.

Additionally, initially, it was not clear which manufacturing control regulations (e.g., GMP, QSR, or both) and which compliance guidance was to be used during the preapproval inspection. Similarly, it was also not clear which adverse event reporting requirements should have been used during investigational development (e.g., 21 CFR 312 or 21 CFR 812).

Since expectations during inspections are undefined and investigator training or experience may not include reviewing combination products, companies may face delays. In some instances, these companies may have to also face duplicative inspections from different centers. Moreover, 483 observations generally focus on lab, quality, and production systems, areas that may be unclear, placing developers at higher risk of noncompliance, possibly resulting in shutdowns and other setbacks. Confusion regarding regulatory requirements and difficulties associated with implementing compliant systems can prevent or slow down developers from pursuing and making innovative products available in a timely manner to the public.

Specific issue-directed inspections

Examples of nonconformities found by agency inspections are listed below:

- The change control system to capture all the changes for the development and manufacture of the combination product was inadequate, as all changes made to the product were not captured and could not be tracked.
- Advertising and promotional materials were not revision controlled or reviewed by the key individuals required to approve the changes.
- Training within a company manufacturing a device-drug combination product had addressed training needs for design controls and QSR; however, training for cGMP and the pharmaceutical requirements and regulations was inadequate.
- Documented procedures for the manufacturing processes addressed the device requirements but failed to address the biologic requirements in a biologic-device combination product.
- Validations performed on new equipment/processes for the manufacture and release of a novel combination product were inadequate.

A compliance program for combination product manufacturers

For combination product manufactures, meeting the compliance requirements is critical in not only launching the novel technology and product in many instances, but also for staying in business over time. When compliance issues are identified within a company by means of vendor audits, internal audits, and so forth, it is extremely important to address these compliance issues in a swift and timely manner to reduce the noncompliance impact to the company and the product. It is also important to bring these issues to the attention of top management and identify the risks associated with the noncompliance issues, so that adequate resources and finances can be allocated within the company to address these needs.

In order to minimize the noncompliance risk faced by a company, a robust compliance program needs to be implemented within an organization. Elements that need to be addressed are listed below:

- Senior management needs to be committed to and support implementation of a compliance program.
- Identify key individuals within an organization to lead and manage the compliance initiative.
- Develop and implement clear document procedures and policies on how to achieve a compliance state within a company routinely.
- Effectively communicate the progress, setbacks, new regulations, and guidelines within the organization.

- Implement training programs to address both sets of regulatory requirements for a combination product. For example, for a drug-device product, training should be provided for both the drug and the device requirements. Similarly for a biologic-device product, training should be provided for both the biologic and the device requirements.
- A strong internal audit program should be implemented to detect and monitor deficiencies earlier within the company, rather than by an external agency.
- Effective supporting systems should be set up, such as risk management, complaint, and corrective/preventive action systems to address and resolve identified nonconformances.

In conclusion, it can be stated that the FDA has various different inspections that can be conducted for single-industry companies, such as medical device, biologic, or pharmaceutical, by their respective divisions, such as CDRH, CBER, or CDER. With combination products, it is not unusual for a manufacturer to undergo both drug and device inspections prior to product approval.

Recent information also suggests that due to the limitations on funding for the FDA, the agency is prompted to shut down regional offices and certain field offices to use its limited resources more efficiently. A streamlined GMP inspection program is proposed with increased responsibilities residing with manufacturers to ensure process and product quality. This streamlined inspection program will be based on identifying drug manufacturing sites most in need of frequent oversight by FDA versus those companies with lesser risks having reduced inspections. Additionally, the FDA's focus would like to change from total number of plant inspections to evaluation and focus on how successfully the FDA can ensure medical product quality and safety.

Agency audit readiness

Combination product manufacturers should plan ahead and prepare their organizations to deal with agency audits and audit-related activities. Many different areas need to be addressed. These include but are not limited to:

- Having procedures and policies for agency inspections that address topics such as receiving the FDA inspector, use of any applicable regulations and standards, establishing roles and responsibilities, preparations to be made prior to an audit, and so on
- Guidelines for acceptable and unacceptable activities and behaviors during an audit
- Defining, establishing, and conducting mock audits
- Dealing with the agency at non-U.S.-based manufacturing locations
- Plans to deal with potential warning letters, FDA 483s, or other outcomes

There are quite a few publications and conference papers that are available to address agency audit readiness. It is our opinion that in order to deal with audits for combination products, it is safe to assume that the manufacturer will have to deal with representatives from various FDA agencies that look into all the constituent parts of the combination products. As such, it is necessary that the manufacturers address the areas mentioned above and, in addition, ensure that appropriate personnel are assembled and trained to answer agency inspectors' questions adequately, so that the inspections are completed successfully.

Bibliography

Application integrity policy procedures. FDA, March 5, 1998.

Application of cGMP regulations to combination products: frequently asked questions Proposed guidance, Office of Combination Products, January 15, 2007.

Axelrod, Michelle D. Is misconduct worth a corporate integrity agreement? *Regulatory Affairs Focus,* January 2007.

Comments on FDA adverse event concept paper. CPC, March 23, 2006.

Current good manufacturing practice for combination products. *Federal Register,* April 24, 2006.

FDA Compliance Program Guidance Manual. Program 7346.843: *Post approval audit inspections.* September 15, 1994.

FDA http://www.fda.gov/cder/dmpq/compliance_guide.htm-top#top *Compliance Program Guidance Manual.* Program 7356.002: Drug manufacturing. February 1, 2002.

FDA Compliance Policy Guide 7150.09, Section 120.100, Fraud, Untrue Statements of material facts, bribery, and illegal gratuities. FDA, July 1, 1991.

Human Drug CGMP Notes, Volume 2, Number 1. FDA, CDER, March 1994.

Michor, Salma. How to avoid non-compliance. *Regulatory Affairs Focus,* pp. 10–13 January 2007.

Office of the inspector general's compliance program guidance for pharmaceutical manufacturers. *Federal Register,* May 5, 2003.

Points to consider for internal reviews and corrective action operating plans. FDA, June 1991.

Portnoy, Stuart and Koepke, Steven. Regulatory strategy: pre-clinical testing of combination product. *Medical Device & Diagnostic Industry,* May 2005

Regulatory Procedures Manual, subchapter 10, Section 10-9, p. 28. FDA, March 2004.

Tetzlaff, R. Validation issues for new drug development: Part III, Systematic audit techniques. *Pharmaceutical Technology,* pp. 80–88, January 1993.

Vinhais, Joseph. Integrating compliance-based solutions for the enterprise. *Regulatory Affairs Focus,* September 2006.

Wechsler, Jill. FDA moves to streamline GMP inspections. *BioPharm International,* pp. 22–24, June 2007.

chapter ten

Conclusions

Any discussion on combination product challenges would be incomplete unless the reader thoroughly understands the necessary market dynamics in medical device, biologics, or pharmaceutical industries.

The medical device industry

It could be argued that the medical device industry, due to less technological barriers compared to its pharmaceutical counterparts, has seen slower product innovation growth whereby medical device products with little to no differentiation compete with each other in the market. A quick analysis of the devices approved in the past few years reveals that most of these devices are either incremental or supplemental changes to the existing products rather than breakthrough creative and innovative technologies. Several key device companies are generating new business and technological opportunities through the integration of devices with other complementary technologies in the pharmaceutical and biologic industries, thereby increasing market share. The diagnostic tools market is also utilizing device components, and additionally, the devices are becoming effective platforms for drug delivery.

The biotechnology industry

The biotechnology industry is in a growth phase at present, with many biotech companies with high intellectual property assets and private funding for customized and innovative biological technologies commanding high prices. In this industry the number of new product submissions to the Food and Drug Administration (FDA) has remained stable. Many companies, as a result, with allocated funding, limited products in the developmental pipeline, and so on, are also looking at mergers and collaborations with other biotech, pharmaceutical, or device firms to develop and launch novel combination products, thereby having a stronger position than they would on their own.

The pharmaceutical industry

The pharmaceutical industry is a mature industry with declining revenue growth and profits, increasing R&D costs, limited duration of patent protection, and so forth. These issues have forced pharmaceutical companies to look for means to increase their innovative growth in novel products, resulting in greater efficiencies and profits. In addition, greater competition from alternatives such as biologic products and generic drugs available at cheaper cost to the consumer is forcing the pharmaceutical industry to assess its current product portfolios and develop novel growth opportunities for these companies. As a result, several pharmaceutical companies are looking at collaborative efforts with device and biologic firms to develop and launch novel combination products. This partnership with other biologic or device companies in turn may also increase profits and reduce development costs.

The chapters presented in this book clearly show that the combination of drugs, devices, and biologics can lead to innovative and financially beneficial healthcare solutions for the future. These combination products are generally complex in nature, with fundamentally different components that work together to achieve a greater medical benefit. Such products have far-reaching applications and include cardiovascular devices, orthopedic devices, wound dressings, tissue implants, organ substitutes, and so on. The market growth for combination products is estimated at about 9 billion by the year 2009. It is also suggested that approximately 30–35% of all new products developed comprise combination products, with devices combined with either drug or biologic products.

As mentioned in Chapters 1 and 2, the combination product groups include diagnostics that provide target drug therapies, devices for monitoring patients and precise delivery of drugs, and medical devices that are either coated, filled, or packaged with biologics or drugs. Introduction of these innovative and novel combination products into the market is in earlier diagnosis, increased patient-centered care, convenience, value, and more effective and safer treatments for many patient conditions. As a result, there has been significant growth in the novel development of various combination products within the medical device, biotech, and pharmaceutical industries, leading to the development and launch of many combination products over the past 5 to 7 years.

Examples of combination products

As discussed in Chapter 2, there are several examples of combination products that have been developed and approved in many cases.

Drug-device combination

Traditional drug delivery systems have consisted of combined or packaged drugs together with injection devices to simplify administration of the product. The list has included such items as pen-based delivery systems, autoinjectors, prefilled syringes, and drug pumps.

Novel drug delivery systems such as patches, transdermal/intradermal injections, inhalation devices, sprays, and drug-eluting disks usually combine existing drugs with new delivery devices. Because their primary therapeutic mode of action is drug related, these products are primarily governed by the regulatory pathway for drugs.

Drug-enhanced devices such as drug-eluting stents, bone cements with antimicrobial agents, coated catheters, and anti-infective sutures are products introduced to enhance the functionality, efficacy, or performance of devices. In many instances, these products combine existing devices and drugs. In these cases, since the primary therapeutic action arises from devices, CDRH primarily governs these products, with a secondary review from drug-related regulatory agencies (CDER).

Biologically active components are combined with devices for regenerative medicinal products that facilitate healing and regeneration of damaged tissues. Examples of these products include Dermagraft (human fibroblast–derived dermal substitute), coated absorbable meshes for bone growth, spinal fusion cages with recombinant human bone morphogenic proteins, and artificial replacement organs (e.g., bioartificial pancreas). These are the most complex combination products because they have to take into consideration the interaction between the product and the body's response to it. The development process for such products is also extremely complicated and integrated since the components under development need to be tightly coupled. The primary mode of action in each case varies on a case-by-case basis, as does the lead review and oversight by respective agency.

Additional examples of various combination products include the following:

Device-drug
- Drug-eluting stent for opening and prevention of restenosis in coronary and peripheral arteries
- Implantable, programmable pump for delivery of a drug or biologic in small, timely doses
- Implantable polymer wafer for release of a chemotherapy agent to a specified site
- Implantable neuromodulator for targeted, regulated delivery of a drug/electrical stimulation
- Transdermal patch for transportation of drugs to local sites or systematically through the skin
- Prefilled, metered-dose syringe, injector pen, or inhaler
- Bone-grafting scaffold/sponge coated with a growth protein that promotes bone regeneration

Diagnostic-device-drug
- Glucose monitor with an insulin pump

Examples of recent developed and introduced combination products include the device and drug combination resulting in the drug-eluting stent (DES), whereby a tiny, bare-metal stent is coated with a drug that would aid in the prevention of restenosis (reclogging of arteries) after heart surgery.

Cordis Corporation, a Johnson & Johnson's device division, launched the original DES, CYPHER, in 2003, achieving worldwide sales of $1.9 billion in 2004 and $2.6 billion in 2005. A competitive DES stent, TAXUS, was launched by Boston Scientific in early 2004, which also achieved significant sales. Recent introductions of drug-eluting stents by Medtronic and Abbot have made this field very competitive. Besides drug-eluting stents for coronary applications, examples of companies developing DESs for peripheral arteries can be seen with Cook Medical, which launched a clinical trial of its peripheral drug-eluting stent ZILVER PTX in 2005, targeting the prevention of restenosis in the arteries that supply blood to vital organs like the kidney and liver.

Another example of a drug delivery device is INFUSE, Medtronic's spinal cage fusion solution that involves implanting a prepacked metal cage with a bone growth–promoting protein. This combination product is designed to replace difficult and painful bone-grafting procedures with a natural bone regeneration process. Strong growth for this product has also attracted other competitors, such as DePuy Spine (J&J).

Yet another example of an implantable drug delivery device is GLIADEL, a small implanted polymer wafer that provides controlled and timely release of a chemotherapy agent in the treatment of malignant brain tumors. The drug minimizes systemic toxicity, as it is delivered only to the tumor site. Gliadel is marketed by MGI Pharma and developed by Guilford Pharmaceuticals.

An example of a diagnostic-drug combination where the integration of diagnostics, drugs, and instrumentation can be seen is used to improve the diagnosis and treatment of, for example, cancer, whereby detection is in the very early stages with these products, such as with positron emission tomography (PET) scanners used in combination with pharmaceutical contrast agents and radiopharmaceutical tracers. Comparison of the biological reactions of healthy and potentially cancerous cells to the radioactive agent, injected into the patient, enables the scanner to diagnose cancers at earlier stages and less invasively than through surgery.

Diagnostic-device-drug

Device implants can provide patient monitoring systems that are less invasive and painful, and more convenient diagnostic and monitoring methods than traditional ones. An example of this can be seen with blood glucose monitors combined with implanted insulin pumps. By integrating convenient monitoring with timely, controlled release of insulin, this combination product provides diabetes patients with a less invasive and more effective treatment alternative. Medtronic first launched its Paradigm Link (512/712 series) glucose monitors in 2004, followed by their MiniMed Paradigm monitors, resulting in significant sales growth for Medtronic.

In our discussions in earlier chapters, the drug-eluting coated stents manufactured by various medical device companies were highlighted as a family of novel cardiology products that spiked interest and led to the release of some regulations (if not all-encompassing) for combination products. Even within this coated-stent market, it is interesting to note that there are currently advances under way to manufacture these drug-coated stents with reabsorbable materials with lesser risk and greater safety to the patients for future use. Such continuous entry of these novel and innovative technologies also poses significant challenges for the FDA in determining suitable policies, processes, and guidances to be developed for industry for greater standardization in approval, release, and postmarket control of these combination products. For the industry and manufacturers of these products, the regulatory pathways of approval as well as manufacturing control of methods, processes, and so on, are extremely challenging in identifying an optimal pathway for any given product, from development through launch.

Development of combination products is recognized by the various medical industries to be extremely important in therapeutic benefits by means of placing these combination products into the market, that is, drug-device or biologic-device product for drug delivery targeted at certain areas of the body in comparison to exposure of the drug to a more general application within the body, resulting potentially in greater toxicity and systemic effect. As scientific advancement gives rise to more and more combination products, the FDA and manufacturers will continue to struggle with the issues raised by regulation of combination products, unless those issues are addressed soon. For these reasons, the FDA needs to work with combination product manufacturers and other stakeholders to develop and implement appropriate solutions rapidly to address these issues. By implementing relevant guidance documents and regulations, the FDA can reduce and eliminate duplication and confusion within the current system for combination products. These changes by the FDA would eventually lead to an effective and efficient system, thereby providing assurance for safe and effective novel products to be available in a timely manner.

Factors involved in combination product development

In Chapter 3, we presented the development challenges and a pathway to develop combination products. The discussions on development approaches taken typically by medical device, pharmaceutical, and biologics industries highlighted that the economic and industry conditions for these healthcare sectors vary significantly in terms of patent expiration, product lifecycle, and so on, whereby each industry faces unique challenges in maintaining its market position and business growth for its relative industry. Each sector is also, in most cases, faced with reduced profits from R&D pipelines, increased competition regardless of the specific industry, financial pressures on cost, and so on. Therefore, combination product development for many such companies in various life science industries provides a route for these companies to develop and market innovative technologically advanced products in an otherwise uncertain industry and environment. The development pathway to combination products has to be hybrid in nature, where appropriate elements of medical device, biologics, or pharmaceutical development processes must be brought together. Without such an approach, it would be extremely difficult or almost impossible to develop these products that can meet or exceed market and regulatory expectations. Concepts presented in Chapter 3 should help the reader in deploying such a hybrid process and associated systems.

The hybrid pathway presented in Chapter 3 also highlighted the need for many areas, including, but not limited to, facilities, personnel, IT systems, and team structure. In Chapter 5, we presented other support system considerations in addition to technological challenges in the development of combination products. With adequate and competent IT resources, technical resources, facilities, and quality, regulatory, clinical, and manufacturing systems in place, manufacturers of combination products can ensure a higher probability of success.

Success for a combination product company

The success of a company developing a combination product will depend on many factors, including defining a successful developmental pathway for the combination product, the company's business priorities and current position in the market, any potential opportunities for partnership with a complementary business collaborative company, and identification of all associated risks with the development and launch of the combination product, which could include resource requirements, facility infrastructure, subject matter expertise for employees involved in the development of the combination product, and regulatory, manufacturing, and quality requirements.

A key challenge in identifying a successful development partner for combination products results in ensuring that the goals, vision, capabilities, knowledge, and expectations for companies entering into the partnership are aligned.

Pitfalls and challenges of developing combination products

As discussed in Chapter 7, there are many different paths a company can take in developing combination products. Additionally, as noted in these chapters, there are significant risks involved. Development of combination products for many companies, both large and small, is a daunting and unfamiliar experience as a result of little to no experience and novel pathways for many companies. Quite often there are a great number of uncertainties related to not only the predevelopment pathway, but also the technological components and integration of a combined product, whether a device-drug or a biologic-device. The question around suitable collaborators is critical to the success of the final combined technology and product launch and, finally, the question of how successful the new technology would be in the market and how well is it accepted postlaunch.

As discussed in Chapter 7, with respect to uncertainties faced by a company in regards to the technology, this could be due to the component uncertainty (whether or not the key components are already available for a combination product) or the integration uncertainty, which refers to whether the necessary interfaces are in existence to integrate the components into a combination product. An example of this can be seen with the drug-coated stent, whereby a specific type of stent was needed for the drug adherence.

Component development

When companies have higher uncertainties on the performance of specific components, the choices are to invest internally in the research and development to develop the component, which can lead to proprietary technologies for the company, or to purchase components developed externally.

Integration of components

In cases where it is a known fact that component uncertainty is low, the effective integration of these components is key. In these cases, companies can increase their in-house efforts to develop unique and proprietary interfaces, or they outsource these to external firms. The company also has another option: to collaborate with another company.

Material selection for combination products

The growth of combination products, together with the evolution of the FDA's regulatory oversight, is forcing these companies to identify materials that could potentially support both performance and compliance needs for combined products.

Consideration for materials for combined products would include not only the mechanical and physical performance needs of applications, such as strength, impact resistance, lubricity, chemical resistance, and stiffness, but also superior material surfaces to reduce interactions between the biologic or drug and the device, biocompatibility, and effects of sterilization on the finished product.

In cases whether an *in vitro* diagnostic, a biopharmaceutical, or combination products that may require protein contact with a plastic device or component would benefit greatly from improved materials with the capacity for nonbinding to low-adhesion surfaces, which helps greater amount of protein to be contained within the biologic.

Manufacturers of combination products also need to consider biocompatibility requirements or issues as a result of device-biologic or device-drug integration.

Combination products are a promising market for healthcare manufacturers. However, the complexity of these products together with increasing regulatory/legislative requirements mandate careful consideration of the materials used. Manufacturers need to evaluate the ability of a resin or compound to help meet requirements such as advanced surface properties, biocompatibility, and food contact compliance, as well as retention of properties following sterilization.

Choice of pathway for a combination product

A company has to identify and develop a strategy on how it can successfully develop a combination product and which pathway it should take to the development and launch of a successful combination product.

Regulatory challenges

As discussed in Chapters 4 and 8, the regulatory requirements for pharmaceuticals, medical devices, and biologics are different, governing premarket applications, design and development, manufacturing requirements, postmarket reporting of adverse events, complaints, and so on.

Combination products quite often present a significant regulatory challenge in terms of which regulations are applicable to individual combination products.

As discussed in Chapter 4, each product type is regulated by a different office within the FDA in the United States. The Office of Combination Products (OCP) was established in 2002 to specifically focus on the development of regulatory guidelines and compliance systems for products that combine two or more product types. All products submitted for the FDA for approval are usually assigned to a particular center within the FDA with primary jurisdiction for regulation of the product. Devices, biologics, and drugs each have their own center (CDRH, CBER, and CDER, respectively). Combination products in the United States are managed through the OCP and classified by the primary mode of action. Additionally, the combination product's components are usually reviewed by respective FDA centers.

As discussed in Chapter 4, in Europe, a combination product is categorized as a medicinal (pharmaceutical) product or a device based on how the product is integrated. Examples can be seen whereby a syringe is regulated under the Medical Devices Directive and a drug administered using the device is regulated under the Medicinal Products Directive. A prefilled syringe would be considered a medicinal product, since, under European regulations, a device and drug that form a single, integrated product are regulated as these. Combination products developed by U.S.-based companies need to be aware of recent and future regulatory changes in Europe and Asia. Whether a single application or multiple applications are submitted to the FDA varies on the specific combination product, the OCP's determination, and the business strategy for these companies and their combination products.

Given the challenges and uncertainties involved in the regulation of combination products, discussions with regulatory agencies early in the product development process are critical for avoiding delays and progress with the product, not to mention incremental costs. Additionally, these continuous discussions with the agency can also ensure any regulatory changes implemented are not missed and can be taken into account in the early stages. Regulatory experience with employees and industries with combination products is relatively new, adding extra time and effort to bringing these products to market.

The FDA is faced with unique challenges for combination products that can arise from various factors, such as policies specified for specific centers within the FDA, different development cycles and timelines, and the complexities of the different types and numbers of combination product configurations.

There are substantial differences in the way the various FDA branches function; for example, the Center for Drug Evaluation and Research (CDER) and the Center for Biologics Evaluation and Research (CBER) are critical in assessing any changes affecting product quality, potency, strength, or purity prior to granting market approval. Any formulation changes to a drug or biologic are usually required to be supported by significant and adequate levels of product testing. With biologic and pharmaceutical products, a key area for collecting data is focused on the information for product stability.

The CDRH branch that focuses on medical devices within the FDA is more focused on the design and development of the product by means of design controls, and in particular whether the risk assessment for the product was adequately performed and manufacturing controls.

A device manufacturer pursuing market approval of a combination product, for example, a device-biologic combination product, may not understand that CBER would have greater expectations and different testing performance standards for the biologic component of the product than CDRH for the device component.

Manufacturing of combination products

Combination products are unique since these manufacturers have a necessity to meet two sets of quality system requirements. These include the quality system regulation (QSR) requirements applicable to medical device manufacturing and current good manufacturing practices (cGMP) requirements applied to drug and biologic manufacturing. To manufacture a combination product, the FDA requires both systems to be considered during and after combining the device and component parts.

In the case of drug manufacturers already meeting the regulatory cGMP requirements, the QSR elements required for consideration are design, purchasing controls, and corrective and preventive actions. In contrast, for device manufacturers already meeting the regulatory requirements for QSR, the cGMP requirements for the drug or biologic that need to be considered are container testing, drug stability, packaging integrity, sterilization effects, biocompatibility, shelf life, and retention of samples.

Packaging challenges

The advent of combination products into the markets has created significant challenges for packaging for the manufacturers of these products. These manufacturers are now forced to consider regulatory requirements, actual materials used in the combination product, equipment, and their effects on the combination product and sterilization methods when packaging these cutting-edge drug-device, device-biologic, or drug-biologic products. This has resulted in a totally different set of packaging requirements and testing for the developer of the combination product. Questions around whether the container and closure system would be suitable for use with the drug would address moisture loss or gain in the package, evaluation of extractables and leachables to and from the package, and product stability requirements. Further drug product packaging requirements would involve batch-to-batch uniformity of materials to ensure product safety and efficacy.

For biologic products used in combination products, most biologics are complex mixtures that are difficult to characterize. These biological components of a combined product are often heat sensitive and susceptible to microbial contamination; it is therefore necessary to use aseptic conditions for these products all through the development and manufacturing stages. These biologic components for a combination product would pose substantial packaging challenges. Biologics often degrade quicker, and hence are time sensitive and more vulnerable than devices due to their easily destructible physical properties. Current packaging challenges associated with combination products, such as delivery of a sterile device now coated with a fragile biologic, raise new concern.

It can be seen that in the case of a medical device company, combination product packaging would present novel challenges since now the design of the package needs to also incorporate the required elements of either pharmaceutical or biologic packaging based on the nature of the combination product. The principles of drug packaging versus device packaging are based on different priorities, whereby in the case of the drug packaging, the primary focus is on safety and effectiveness of the product; in comparison, device packaging focuses more on protecting the product, allowing and maintaining sterility for these devices if they are to be used sterile.

For a device-drug combination product, the packaging system should adequately protect the finished product from exposure to light, reactive gases like oxygen, water vapor absorption, contamination, and so on, which may alter the drug dosage.

Sterilization

Sterilization is another key area that needs to be considered for finished combination products. The combination of drug, biologic, and device forces the manufacturer to provide careful consideration on which sterilization technique will be used for the respective combination product, since each element of the combination product may respond in a different manner to the various sterilization techniques. For example, ethylene oxide (EtO) sterilization of liquids is not common; additionally, gamma irradiation of biologics or pharmaceuticals is not typical, and furthermore, many resins currently available are also not compatible with EtO, gamma, or steam autoclave without consequences to the finished product.

Since the efficacy of drug or biologic molecules is very likely to be dose dependent, every effort needs to be made with these fragile components of a combination product that the lowest possible radiation dose required to make the products safe and effective for their intended use is utilized.

The method of sterilization oftentimes complicates the timelines, successful development, and final approval of combination products. An example of this can be seen with a device-pharmaceutical combination product, where ethylene oxide is commonly used to sterilize medical devices; however, ethylene oxide reacts when it chemically interacts with the drug substance. This results in the pharmaceutical component to have reduced effectiveness, potency, and so on. The sterilization method is also important in how the determination for a given lot will be determined for a combination product.

Additionally, sterilization is more complicated with combination products. The manufacturer needs to not only be sure that all components within, for example, a combination product kit reach a sterile state, but also ensure that all the materials used are compatible with the sterilization process. In many cases, the package design of the final product is often expected to facilitate sterilization. Packaging design has to be considered seriously if the device manufacturer employs gamma irradiation to sterilize its products. When utilizing irradiation for sterilization, both product and packaging materials have to be accounted for in any nonadverse impact by the sterilization process. In certain cases, an incorrect material could result in brittleness or discoloration following sterilization.

Sterilization practices may have to change when a combination product is involved. Device manufacturers have to understand what effect their sterilization methods have on drugs or biologics. Usually, biologics and drugs are not subject to terminal sterilization like devices. Therefore, when terminal sterilization is planned, the materials used to create packaging or devices need to be appropriately chosen to allow for the sterilization conditions of both the drug and biologic as well as the resin materials themselves. The chosen sterilization process also needs to retain the mechanical and physical performance of the package or device poststerilization. Biologics and drugs are most commonly manufactured in a clean room or by aseptic processing.

In conclusion, medical device, pharmaceutical, and biologic companies developing and manufacturing combination products need to appreciate the significant differences between CDRH, CDER, and CBER when obtaining FDA approval for their respective products. The FDA's requirement for drug and biologic testing is more stringent and complex than its requirements for medical devices. As long as the product under review from the FDA allows a clear classification as a device, drug, or biologic, the regulations provide clear guidance on which of the three centers would conduct the evaluation. However, with a product relying on different modes of action, the agency's process for determining the regulatory path to be followed is not always predictable.

Recovering developmental and manufacturing costs is also challenging for companies developing combination products, in addition to the regulatory challenges faced by these companies in bringing novel technologies to the market. In many instances, a particular technology to bring a specific combination product may have excessive costs associated with it, due to either the complexity of the technology or the manufacturing challenges or reimbursement issues once the product is released. These innovative medical technologies with combination products would likely incur high costs during development, regulatory review, clinical trials if needed, and manufacturing.

Manufacturing of combination products is also significantly higher due to not only the increased complexity of the manufacturing processes and equipment, but also the additional cost of meeting compliance requirements to regulations across multiple product categories, such as device, pharmaceutical, or biologic. Additional costs are also incurred due to extensive labeling requirements for combination products, shelf life testing of these products in order to determine a realistic expiration date, and so forth.

In regards to coated combination products, for example, a current device product with a biologic or drug in the case of the drug-coated stents, coating of the stent was looked at as being a benefit to improved efficacy and significant profits. Some of the key challenges faced by companies introducing the drug-coated stents were initially a regulatory concern of which FDA center would be primarily responsible for product regulation. This was magnified by the fact that most device companies have little to no experience with the drug regulations or technology used to coat the device. Key questions that needed to be addressed for these companies were what drug to use, would the drug require modification, and what technology would be used to incorporate the drug onto the device. Additionally, unlike devices alone, coated combination products such as the drug-coated stents needed to have toxicity levels determined, characterization of the drug's release into the body at the site, and an acceptable shelf life. These products also had to undergo preclinical and clinical studies to show safety and effectiveness in the human body for its intended use.

There are significant coated combination products that the CDRH has regulated as devices in the market. Some of these products include antimicrobial bone cements, intravascular and urinary catheters, and heparinized intravascular catheters.

Some key questions to be addressed by companies developing combination products are:

1. Should partnerships be initiated at the R&D, manufacturing, or marketing stage?
2. In regards to technology, do the necessary components, interfaces, and integration capabilities exist, or do they need to be developed?
3. Does the company have any of the core competencies currently related to development of a combination product?
4. What are the core competencies a company needs to develop, produce, and launch the new product?
5. Are there other external companies with these core competencies?
6. Can the existing R&D processes, manufacturing practices, information systems, facilities, and other operational elements support the introduction of a combination product, or would they need to be altered to support the development and production of the combination product?
7. Are the current organizational structures sufficient in terms of expertise, or do they need to be restructured to support development of a combination product?
8. If new skills and specialized expertise are required, how will they be obtained? If there are collaborative partners, will the cultural differences between the companies be aligned or cause additional problems?
9. Does the company have adequate resources for building, borrowing, or buying the capabilities it needs?

As the combination product strategy is implemented, it is critical to revisit these questions to consider whether directional adjustments are needed.

In general, many companies that have developed combination products in the last several years, rather than developing in-house capabilities, have collaborated with companies that complement their requirements in securing the components, interfaces, and integration capabilities needed. By partnering with other companies, these specific organizations have obtained technologies already available, reducing the time required, and accessed expertise, knowledge, resources, and distribution channels.

As medical device, pharmaceutical, and biologic form collaborative relationships in developing and manufacturing a combination product, they need to recognize the differences that exist in these various industries. A key element of success is for these different industries to establish clear goals, objectives, expectations, and responsibilities for their partnerships. Strong interactive meetings, joint planning, working to a single product development strategy derived by input from both partners, establishing cross-functional teams, and supporting strong relationship building between the two collaborative partners within their respective organizations are key. Additionally, leveraging of expertise between the two industries would enable cross-training and sharing of knowledge and different skill sets.

We strongly believe that we have provided a comprehensive look at various aspects of combination product development and regulatory challenges. When properly utilized, the approaches we presented in this book should help companies developing combination products to successfully overcome these challenges.

Bibliography

21 CFR, Part 210, 2006.

21 CFR, Part 211, 2006.

21 CFR, Part 610, 2006.

21 CFR, Part 820, 2006.

Barrett, Amy. Unclogging J&J's pipeline. *Business Week Online*, March 20, 2007.

Butschli, Jim. Healthcare packaging: combination products challenge packaging. *Packaging World Magazine*, p. 36, June 2006.

Container closure systems for packaging human drugs and biologics. Guidance for industry, CDER, 1999.

Cook launches web site for patients seeking information about peripheral arterial disease, April 23, 2007: http://www.cookmedical.com/newsDetail

Eselius, Laura, Nimmagadda, Mohan, Kambil, Ajit, Hisey, R. T. (Terry) and Rhodes, John. *Managing Pathways to Convergence in the Life Sciences Industry*. Deloitte research study, May 2007.

FDA approves new wafer to treat brain cancer. *Wall Street Journal*, September 25, 1996.

Frissora, Clare. Top healthcare-related material considerations for combination products. *Medical Design Technology*, July 2007.

Loob, William. Combination products enhance capabilities, pose new challenges. *Medical Device & Diagnostic Industry*, May 2000.

Masefield, J. and Brinston, R. Manufacturing: radiation sterilization of advanced drug-device combination products. *Medical Device Technology*, March/April 2007.

Portnoy, Stuart and Koepke, Steven. Regulatory strategy: pre-clinical testing of combination products. *Medical Device & Diagnostic Industry*, June 2005.

Q1A(R2) stability testing of new drug substances and products. Guidance for industry, CDER, 2003.

Richter, Steven. Combination products: navigating two FDA quality systems. Microtest, white paper, January 2007.

Sullivan, John and Elbaek, Henrik. Combination products face additional regulatory hurdles abroad. *Medical Product Outsourcing*, April 2007.

Swain, Eric. Extra effort: packaging for combination products. *Medical Device & Diagnostic Industry*, January 2005.

Triolo, Phil. Coated combination products: regulation and technology. *Medical Device & Diagnostic Industry*, March 2005.

Viscogliosi, Marc R. Building a new model for spine care: when the dust settles, technologies and procedures for treating spine conditions will look very different. *Medical Device Link*, May 2007.

Index

H

I

J

L

M